Superdistribution

Objects as Property on the Electronic Frontier

Brad Cox

George Mason University Program for Social and
Organizational Learning

ADDISON-WESLEY PUBLISHING COMPANY

Reading, Massachusetts · Menlo Park, California · New York
Don Mills, Ontario · Harlow, United Kingdom · Amsterdam · Bonn
Sydney · Singapore · Tokyo · Madrid · San Juan · Milan · Paris

Thomas Stone *Sponsoring Editor*
Katherine Harutunian *Associate Editor*
Helen Wythe *Senior Production Editor*
Judy Sullivan *Manufacturing Coordinator*
Barbara Atkinson *Cover Designer*
John Gamache *Cover Illustrator*
The Beacon Group *Composition*

Library of Congress Cataloging-in-Publication Data

Cox, Brad J.
    Superdistribution : objects as property on the electronic frontier
  / Brad Cox.
       p.   cm.
    Includes index.
    ISBN 0-201-50208-9
    1. Computer software.   2. Data transmission systems.
  3. Information superhighway.   I. Title.
  QA76.754.C69   1996
  338.4'7005--dc20   95-51747
                                                                 CIP

Many of the designations used by manufacturers and sellers to distinguish their products are claimed as trademarks. Where those designations appear in this book, and Addison-Wesley was aware of a trademark claim, the designations have been printed in initial caps or all caps

Access the latest information about Addison-Wesley books from our World Wide Web page: http://www.aw.com/cseng/

1 2 3 4 5 6 7 8 9 10-MA-99989796

Preface

Why spend $2.95 a minute to talk to your psychic adviser when you can be enlightened for free? Why spend hairsplitting weeks waiting for Ann and Abby to read your letter when you can get answers right now? The Net is a vast fountain overflowing with knowledge. So is Allison's head. But where on the Net can you ask how to break off an engagement? Nowhere. Until now.... There's no need to face life's mysteries alone. Allison, our smart and sassy advice columnist, is here to help. In her own words, "Ask me, I know."
— Ask Allison, She Knows NEWS OF THE WIRED

The relationship between the external world of the senses and the internal world of the mind has been the subject of philosophical speculation and dispute since antiquity. No doubt the wrangling will continue ever hence.

Quite recently, within our own generation, this age-old dispute became far more complicated. When computers and telecommunications emerged after World War II, they brought with them an entirely new world, similar to yet unlike the two worlds of antiquity. The entities in this new world are unlike the tangible things of the world of the senses, but neither are they entirely like the intangible ideas of the mind. They are electronic entities in an electronic world. Made entirely of bits instead of atoms, they are so ephemeral yet so concrete that they blur confusingly into both of the other domains.

These entities include computer software in particular and electronic objects in general, the electronic goods of the information age. I will call this new world the *Electronic Frontier*[1] because of striking similarities between this new world and the Wild West of the American Frontier.

TAMING THE ELECTRONIC FRONTIER

The electronic frontier is still very young, less than a single generation. It is still in the pony express stage of the American West. The pony express was a *communication infrastructure*, as are the computer networks we have today. The pony express was capable of hauling only letters and invoices, just as computer networks are capable of hauling only electronic mail, netnews, and web pages.

---

[1] This term was pioneered by the Electronic Frontier Foundation.

Commerce infrastructures capable of hauling passengers and property in addition to just information had to wait for the canals, railroads, highways, and airports. Of course, barring the Star Trek "Beam me up, Scotty" teleporters, computer networks will neither haul passengers nor the other tangible goods of the industrial age. But they can certainly haul electronic goods: electronic property such as the computer software and electronic data that are fast becoming a staple of the emergent information age.

However, this notion of electronic property is still an alien thought today. We still think of computer networks as communication infrastructures exactly like the pony express. We think of E-mail and netnews as *information*. We reserve thoughts of electronic property, such as web pages and computer software, for a very different space in our minds. This other place is inextricably bound up in industrial-age objects such as paper and enterprises such as publishing. The thought that we might use today's communication infrastructures to haul and sell goods is still as foreign as the thought of cross-continental mass transportation of cattle, ore, and coal would have been alien in pony express days.

ELECTRONIC PROPERTY

Electronic property is my umbrella term for the fastest-growing, and arguably soon dominant, segments of today's global economy. Computer software is the example that most of us recognize most readily as property, even though it is almost never packaged that way for sale today. I am also referring to examples that only programmers may recognize, such as reusable software components. But most of all, I am referring to a vast new world of potential examples that are not yet actual because they have not yet been born. These won't be born until the electronic frontier transcends its pony express stage.

Paradoxically, shrinkwrapped computer applications are not electronic property as I'll use that term in this book. They are bought and sold today as tangible goods attached to plastic, paper, and cellophane that can be bought and sold as tangible goods in an industrial-age store. Clip art is another potential example that is also bought and sold as tangible properties, on shrinkwrapped disks and CDs. With only rare exceptions such as the shareware market, software is not marketed as electronic property today, as goods made entirely of bits.

The purest examples of electronic property today are the electronic publications that are beginning to appear on the web. There is a vast number of other potential examples that remain nascent today. This is an outcome of the absence of commercially robust electronic infrastructures that support robust ownership in electronic property. The often sought, but never delivered, dream of a robust market in reusable software components, or Software-ICs, is only one example of many innovations that won't be viable until we move beyond the pony express stage of the electronic frontier. Since computer software and clip art are still bought and sold as hard-to-copy atoms, they are industrial-age goods, not information age goods. Until computers and networks support ownership of goods that are composed exclusively of bits, the electronic frontier will remain untamed.

Existing solutions, such as internet, Compuserve, and America Online, are already fully capable of transporting electronic goods. But these infrastructures neglect what is today the part of the problem that the railroads never had to address. The railroads could concentrate entirely on transporting goods and leave the problematics of buying, selling, and owning them to others. But on the electronic frontier, transportation is no longer the problem. Our problem is the absence of a robust way of buying, selling, and owning property that can be replicated and transported in nanoseconds.

Just as long-haul transportation would have seemed insurmountable in pony express days, our established paradigm is that this ownership problem is insurmountable. Overcoming this paradigm, and the much smaller task of actually building and deploying a practical solution, is today just as formidable as building railroads across the continent must have seemed in its day.

This task is not primarily a techno-centric problem of wiring together a "Nationwide Information Infrastructure." There are lots of serviceable communication infrastructures already, and these will improve as the demand for them grows. Nor is building stuff a particularly big problem for the computer industry these days. In this case the stuff that remains to be built is not a communication mechanism but an ownership mechanism, something that can play the role that commercial exchange transactions played in human affairs since antiquity.

As I'll show in this book, the missing technology is not particularly complicated. The computer industry routinely builds far more complicated technology every day. The problem is that the solution actually involves our paradigms as well as our tools, our society's deepest beliefs as to what it *means*, and what it *should* mean, to say that we buy, sell, or own goods made of bits. Just as in the wild west, taming this electronic frontier involves a paradigm shift, a change in society's deepest beliefs about what it means to say we *own* a piece of electronic property. This paradigm shift as in the meaning of owning land is fully camparable to the paradigm that fueled the cowboy versus Indian and rancher versus farmer confrontations in the wild west.

FRONTIER CONFRONTATIONS

The wild west was once dominated by freedom-loving explorers and Indian tribes. But the railroads brought in a flood of hard-working settlers and property-conscious merchants that ultimately displaced the former group to the margins. Although the winners of this confrontation called this displacement the "Taming of the West," the displaced groups, of course, had entirely different views as to whether this was an improvement.

Frontiers are boundaries, places where a known world abuts a world that is yet unknown. Columbus' "discovery" of America[2] juxtaposed the familiar Old World with a New World that was unknown to the Europeans of that day.

- To scientists and explorers, it meant confirmation of an abstract theory, but one that had extremely practical implications to key sponsors such as Queen Isabella of Spain. The theory was that if the world was indeed round, India (to the east) could be reached by sailing west.

- To the indigenous people who had discovered America long before Columbus "discovered" it, the American Indians, whose name derived from Columbus' mistake, it initially meant a chance to welcome, entertain, and feed another lost shipload of European sailors.

- To the ordinary citizens of a war-torn, disease-ridden, religiously intolerant Europe, with limited horizons for all but the nobility, it meant a way any civilian could individually grow and prosper.

Since the European belief that land was something to be owned was anathema to the Indian religious beliefs, it appeared to them that the wealth of this vast continent was free for the taking. So take it they did, pushing entire indigenous civilizations to the very brink of extinction.

The same cultural confrontation is now starting on the electronic frontier. Although this confrontation will most likely avoid at least most of the bloodshed, it will involve confrontations between equally cherished cultural beliefs and the eventual displacement of cultures. This confrontation involves the same set of cherished beliefs that brought the Europeans into conflict with the Indians. The American Indians abhorred the idea that individuals

---

[2] I'll occasionally use the tale of Columbus' "discovery" of America to typify other frontier discoveries of the past. Regardless of the interest and importance of such historical questions as whether Columbus was the first of the European discoverers, and whether he discovered the Indians or vice versa, I'll avoid such questions, as well as moral judgments about which culture's belief-system is "right" in absolute terms. This would be a needless digression and take us into areas beyond my competence.

could own land, whereas the Europeans believed exactly the opposite. In our case, the cherished belief in question is the deeply established belief that property made of bits is a contradiction in terms, that "information wants to be free."

This book originates from the viewpoint of a member of the electronic frontier's indigenous tribe, the "nerds." From the vantage point of my career in the software industry and my subsequent roost in academia's ivory tower, I've seen a horde of strange invaders, the "newbies," just over the far horizon. This book reports on where this invasion is leading and what it may mean to both sides in the end.

I will leave ethical arguments as to which side of this confrontation is morally "right," in any absolute sense, to those with far more wisdom than I. My role is to warn that as with the Indians in North and South America, the Maoris in New Zealand, the Aborigines in Australia, the Ewe in Africa, and other confrontations of history, the communitarian ideal has not fared well in confrontations with European-style cultures.

Although the past is never an infallible guide to the future, it is a guide that only the foolish would ignore. Therefore the task of this book is to show why the European approach is so powerful that it tends to obliterate opposing approaches, why this approach is not yet viable with respect to electronic property, and how this viability may yet be sustained in time.

OBJECT TECHNOLOGY

The origins of this no doubt surprising way of thinking about the electronic frontier are based in early work on object technology. Object technology arose out of early ways of thinking about computer software that denied that there was the slightest similarity between the objects of everyday experience and entities, such as web pages, that reside on networks or the computer software inside digital computers.

The earlier thinking was that, quite unlike the hybrid behavior of everyday physical objects, electronic property had to take either one of two forms. On the one hand was the dead, hyper-passive entities that we called *data*. On the other was the purely hyper-active entities that we called *software*. Object technology arose out of the realization that neither of these bore the slightest resemblance to the tangible objects of everyday human experience. Everyday objects maintain their own state and exhibit their own behavior without the least regard for the historical, and utterly artificial, distinction between 'program' and 'data'.

The recent popularity of object technology displaced the bipolar data versus procedure view with a new world of *hybrids*. Each electronic object can be as active and as passive as that object needs to be, without having to fit into the binary, either-data-or-procedure, mold of traditional software engineering. This movement allowed the reasoning skills that we use for understanding the objects of everyday experience to apply, in a limited but important sense, to understanding the electronic objects that reside on computers and networks. The significance of the term 'object' is apparent in the term itself. It simply means that computers need not be viewed as an alien word of bits and bytes, populated by passive entities called data and active entities called programs. It is a world of *hybrids*, electronic entities that maintain their own state and exhibit their own behaviors, like the tangible objects of everyday experience. It is a world in which the common sense realities of everyday experience have overriding significance to the people who provide and use this new product of human endeavor I'll refer to in this book as electronic property.

My career has been as an active contributor to a movement that ultimately achieved the runaway bandwagon stage, a stage that old-timers recognize as the fate of all worthwhile advances in today's turbo-charged software industry. My earlier book, *Object-oriented Programming; An Evolutionary Approach*, introduced this new technology to programmers in the trenches, who were as a rule using traditional programming languages such as C. It showed that even traditional languages could be extended relatively easily to produce entirely new hybrid languages, such as Objective-C, that support the new features of a new style of software development known as *object-oriented programming*.

In addition to explaining the technical terminology for this new approach to software development, this book introduced a totally new term, *Software-IC* that had not been in software lexicon to that date. I introduced this term to emphasize a key benefit of the new approach. Software-ICs are snap-together software components, that are reusable in ways that preceding reusability technologies (subroutine and macro libraries) were not. The new term was a way of emphasizing that the most radical contribution of object technology was to allow software to be assembled from prefabricated components, rather than by continually fabricating everything from first principles as we had done in the past.

HUMAN-CENTRIC PERSPECTIVE

This book is the outcome of a decade of promoting object technology as a means of achieving what the manufacturing community calls 'specialization of labor' and which the software community calls 'software reuse'. In particular, it is a result of a determined effort to build a commercially robust market in prefabricated software components. This book is for those who are interested in the broader issues that will affect every programmer's livelihood in the future. It is *not* an introduction to object-oriented technologies.

This is the outcome of years of struggle to understand, and then explain, where this latest of bandwagons may take us in the end. I ultimately realized that it would be necessary to adopt a much broader viewpoint than the object-oriented programming community has been using to date. Instead of the techno-centric focus of my earlier work, this one will concentrate on the problematics introduced by objects as a new form of property, goods like cornflakes or automobiles, except that they are composed of bits instead of the atoms of property. I will call this viewpoint a *human-centric* perspective to distinguish it from the *techno-centric* viewpoint of the technical circles I've worked in to date.

The human-centric perspective is reflected in the newbie/nerd conflict with which I opened this preface. It is also reflected in the unconventional way I'll use the term 'object' in this book, as reflected the slogan *Object-oriented technologies are for programmers. Object technology is for everyone.* Whereas techno-centric circles use elaborate *exclusionary* definitions for what can legitimately be considered an object, I'll take exactly the opposite approach here. I will define the term, 'object', without further apology, in an *inclusionary* manner, to include as many narrow technical definitions as possible.

The inclusionary definition of object that I'll use throughout this book is simply *Objects are somebody's electronic property.* This definition implies no restrictions as to granularity, just as the concept of property applies equally to the burger as purchased at the fast food stand, the bun as produced by the baker, the flour as produced by the miller, and the wheat as grown by the farmer. Just as in the tangible world, the domain of discourse will cover electronic objects of all granularities, from large objects like web pages and shrinkwrapped word processor software, to the clip art and reusable software components from which they are composed. This means that the topic will range broadly, from the tiny objects of the object-oriented programming community to the much larger objects that the rest of the world thinks of when "software" is mentioned.

PROBLEMATICS OF OBJECTS MADE OF BITS

In view of the hoopla that has grown up around object technologies, I don't think it is necessary to contribute further to the chorus proclaiming that object technologies really do work. Object-oriented fabrication technologies such as Ada and C++ have proven successful in allowing programmers to fabricate software more productively than ever before. Object-oriented assembly technologies such as Smalltalk and Objective-C have shown that it is also possible to build loosely coupled prefabricated software components that customers can snap together to build finished solutions, exactly as silicon chips are used in hardware manufacturing. Yet our enthusiasm for the strength of a new technology for binding components to each other has become part of the problem. It has caused us to overlook the more fundamental questions of "where will quality components come from? Where is the incentive to provide them?"

Mankind became very proficient at life in the industrial age. We are now moving into an information age in which the objects are no longer tangible entities that can be handled, counted, and weighed. The intangibility causes grave organizational problems wherever they're involved. With respect to the objects produced in the software development industry, they are generally known as the software crisis.

These same problems are emerging under various names, such as 'the white collar productivity problem', at many other levels of granularity too numerous to mention. As the entire global population evolves from an industrial age into an information age, more and more of us become exposed to the very problem I want to engage in this book: What does it mean, or more properly, what *should* it mean to speak of buying or selling this new kind of object that can be replicated and transported in nanoseconds.

To appreciate the magnitude of the issue before us, consider the number of conferences and workshops at which we gather to extol the rapidly emerging new technologies. Then contrast this with the number of conferences that accept these technologies as just another tool in the toolkit, an evolutionary extension to everything we've learned in the past, and move on to paradigm-shifting questions like these:

- Why are libraries of prefabricated 'reusable' software components still the exception and not the rule, even though they've been technically feasible since the subroutine call? Is the problem really demand-side issues, such as inadequate browsers, databases, classification schemes, and the infamous 'not invented here' syndrome? What about supply-side issues such as the absence of commercially robust ways of motivating people to provide components that are designed, implemented, tested, and documented well enough that anyone would want to use them?

- Why is the object-oriented programming language community fragmented into so many warring factions? Why do advocates of assembly technologies and fabrication technologies see diversity as an invitation to language wars, rather than a common sense way of cooperating through specialization of labor?

- Why do some see inheritance as the main contribution of object-oriented technology, whereas others see the main contribution being encapsulation as a means of packaging concrete, prefabricated, pretested software components?

- Why do we continue to measure software innovation in terms of implementation technologies and treat specification and testing as an afterthought? Mature disciplines such as carpentry give at least as much emphasis to carpenter squares and measuring tapes as they do to implementation technologies such as saws and hammers. Why doesn't the software industry ever get around to focusing attention on tools comparable to calipers, rulers, protractors, and inspection gauges, tools that those who build homes and automobiles view as absolutely indispensable?

- Why can't we bring ourselves to speak of selling prefabricated, pretested software components that quality-conscious engineers might buy, as distinct from waste products that we hope trash-pickers might 'reuse'? Why do we lean toward passive words such as 'repositories' when we speak of software components libraries, words that suggest a dank stagnant lagoon where we throw useless stuff to rot? Why do active words such as 'markets', words that successful domains use to mediate their cooperative activities, stick in our throats in connection with information-age commodities like software components?

- How shall we quantize the value of information-age commodities such as software? We already know that it costs more to build high-quality software that others are willing to use. How shall we recover this investment? How will we quantize the value of such components, let alone establish a price for each quantum? Since we can't measure software by the ton or by the barrel, what unit of measurement should we use instead? Should we sell software by the line of source code, or by the method? By the class, or by the instance? Or should we give up on such pay-per-copy schemes as a vestige of manufacturing-age thinking and begin searching for a new basis for commercial interchange in the information age?

- In view of the emphasis put on integrating data and procedure by object technologies, does it really make sense to buy software and data through completely independent schemes, selling software by the copy and data access by the minute in the form of network connect charges? As computers become increasingly networked, might answers be found in using networks to base software value on the usage it receives, not on acquisition of copies?

ACKNOWLEDGMENTS

I would like to thank David Sharp of Taligent, Inc., Martin Haeberli of Netscape Communications Corporation, and Mark Addelson of George Mason University, for their helpful comments as the review committee for this book.

Contents

1 Information Revolution 1

1.1 Computers, Communications, and Mankind 2

1.2 Invisible Plumbing 6

1.3 Breadth Versus Depth 8

1.4 The Electronic Frontier 10

1.5 The Largest Upheaval of All 12

1.6 Taming the Electronic Frontier 13

1.7 Indigenous Encounters on the Electronic Frontier 17

2 Structure of Production 21

2.1 The Tale of the Wooden Pencil 23

2.2 Structure of Production 26

2.3 Social Binding Forces 29

2.4 Commercial Exchange Transactions 30

2.5 The Lair of the Software Werewolf 31

2.6 Pencils as the Fruit of the Pencil Tree 33

2.7 Indian Pottery 34

2.8 Electronic Pencils 39

2.9 Software ICs 41

2.10 Summary 43

3 Software Crisis 45

3.1 The Good News: There is a Silver Bullet 45

3.2 The Bad News: Silver Bullets are Paradigm Shifts, Not Tools 47

3.3 The Copernican Revolution 48

3.4 Software Complexity 51

3.5 Software Industrial Revolution 53

3.6 The Intangibility Imperative 54

3.7 Process-Centric Versus Product-Centric 58

3.8 The Structure of Scientific Revolutions 60

3.9 Software Industrial Revolution 63

3.10 Essence Versus Accidents of Software Engineering 65

3.11 The Four Obstacles 67

3.12 Robust Economics for Information-Age Goods 71

3.13 Summary 72

4 Software Architecture 75

4.1   What, If Anything, is an Object 75

4.2   Heterogeneity 77

4.3   Compositional Architectures 79

4.4   System 12 82

4.5   Fabrication and Assembly 84

4.6   Specialization of Labor 85

4.7   Process Versus Product 89

4.8   Encapsulation and the Law of Proximity 91

4.9   Programming Languages and Operating Systems 93

4.10 A Richer Set of Architectural Distinctions 94

4.11 Five Architectural Levels by Example 97

4.12 Technical Definitions 102

    4.12.1 Process-level Objects 102

    4.12.2 Task-level Objects 104

    4.12.3 Chip-level Objects 105

    4.12.4 Block- and Gate-level Objects 105

4.13 Programming Language Comparison 106

4.14 Summary 108

5 Industrial Revolution 111

5.1   Specification, Testing, and Language 112

5.2   Specification and Testing Tools 113

5.3   Armory Practice 115

5.4   Thomas Jefferson 117

5.5   Eli Whitney 118

5.6   Roswell Lee 123

5.7   Revolutions Don't Happen Overnight 124

5.8   The Heroes of the Industrial Revolution 125

5.9   Software Engineering 127

5.10 Specification and Testing Tools 128

5.11 What is a Stack? 130

5.12 Assertion Checking 132

5.13 White-Box Testing 133

5.14 Black-Box Testing 134

5.15 Black-Box Specification/Testing 136

5.16 Classes and Inheritance 138

6 Out of the Crisis 143

6.1 Out of the Crisis 144

6.2 Copyright and Useright 146

6.3 ASCAP and BMI 148

6.4 Electronic Goods 150

6.5 Large-Granularity Solutions 152

6.6 Royalties 153

6.7 Usage-Based Solutions 154

6.8 Superdistribution 155

6.9 Superdistribution as a Market Mechanism 162

6.10 Superdistribution and Object Technologies 164

7 Commerce Infrastructures 167

7.1 Commerce Infrastructures 168

7.2 User Interface 170

7.3 User Registration 173

7.4 Invocation Metering 174

7.5 Query and Commit 175

7.6 Who Pays the Tab? 176

7.7 Terms, Conditions, and Prices 177

7.8 Tamper-Resistance 178

7.9 Terms and Conditions 181

7.10 Rules of Fair Trade 183

7.11 Registering a Product 185

7.12 Terms and Conditions Algorithms 187

7.13 Financial Institutions 187

7.14 Summary 189

8 Conclusions 191

8.1   The Electronic Frontier 191

8.2   A Personal Retrospective 192

8.3   Will Things Really Develop This Way? 194

Index   195

Chapter 1

Information Revolution

"Imagine discovering a continent so vast that it may have no other side. Imagine a new world with more resources than all our future greed might exhaust, more opportunities than there will ever be entrepreneurs enough to exploit, and a peculiar kind of real estate which expands with development."
—John Perry Barlow, *Decrypting the Puzzle Palace*

Organizations are like fish with people as their cells. They evolved to thrive in the ocean, the high-viscosity world of the industrial age. These fish must now change into fowl to thrive in the zero-viscosity world of the information age in which space and time have collapsed to almost nothing. Most of them won't make it, for evolution doesn't work that way.

Two centuries after its birth in the industrial revolution, the industrial age has reached a zenith and is beginning to show signs of decline. A new age, the age of information, is emerging, born of the phenomenal achievements that the industrial revolution initially brought to manufacturing and transportation, and ultimately to communication and computing. Notice that, in this view, today's computer and telecommunication industries are industrial-age enterprises, not information-age enterprises. They only laid the foundation for the emergence of a true information-age economy at the culmination of the process I'll call the taming of the electronic frontier.

The information revolution is now producing a low-viscosity global economy in which information is becoming as crucial to our prosperity as the tangible things of the industrial age. The televised images of industrial-age bombs gliding unerringly through windows and air shafts during the war in Kuwait is only one of many compelling demonstrations that industrial-age technologies are no longer sufficient and that the information age is with us to stay.

Although most would agree that something new, important, and maybe even frightening is happening, there is still no agreement as to what the information age actually *means*. Although the visionary frontiersmen and women who've now been exploring the electronic frontier for almost a generation are persuaded of the value, this is by no means true of pragmatic settlers and businessmen who are beginning to appear on the horizon, and who lack our enthusiasm for technological whiz bangs.

Regardless of the widely held belief that change is progressing faster today than in earlier periods of human history, this meaning-making activity is a human-centric, not techno-centric, affair. It occurs in the minds and deeds of millions of flesh and blood individuals. Although we're almost never bold enough to admit it explicitly, this meaning-making very often boils down to a new angle for paying the mortgage more effectively than whatever angle we're using at the moment.

Although individuals can in principle adapt quickly to changes in their economic environment, like the cells of a fish that must evolve into fowl, we're enmeshed in a matrix of organizations, institutions, and cultures. These can be remarkably resistant to change, at least in the short run which as a rule means less than a generation or so. Max Planck made this same claim in his famous observation about the history of scientific revolutions, *"Things never change faster than it takes the old generation to die off."*

This chapter will bring into focus an aspect of the information age that is only dimly recognized as a crucial issue today, but which stands directly in the path of the broad acceptance of information technology. The lack of recognition is not because the issue is obscure and hard to understand. It is concealed only by its utter obviousness. It becomes most apparent when you look at the issue from the viewpoint of the pragmatic people at the ends of the wires. I'll call this a human-centric perspective to distinguish it from the techno-centric worldview that has dominated computer science and software engineering since its inception, and which still dominates it to this day.

The tangible goods of the manufacturing age were hard to copy and transport, but trivial to buy, sell, and own. Thus vast commercial enterprises evolved to provide them. By contrast, the intangible goods of the information age can be replicated without loss and transported at literally the speed of light. But we've not yet figured out what it should mean to buy, sell, and own them, which undercuts the commercial incentive to provide them. So the big question of the information age is, if the supply of high quality goods is allowed to remain in doubt, do we really need better infrastructures to haul them?

1.1 COMPUTERS, COMMUNICATIONS, AND MANKIND

Koji Kobayashi, the head of Nippon Electric Corporation, gave the keynote address at the first computer conference I ever attended. His speech, *CC&M; Computers, Communication and Mankind*, held an unusually simple message for a technical conference (Fig. 1.1). He argued that technology would

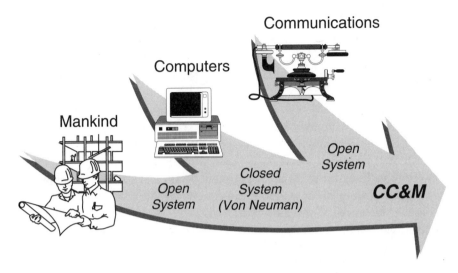

Figure 1.1 Koji Kobayashi, the CEO of Nippon Electric Corporation, predicted that technological advancement would proceed through CC&M, the integration of computers, communications, and mankind. The first phase of this vision has already occurred in the personal computing revolution. The next phase, integrating communications, is still in its infancy.

progress throughout this century through progressively tighter integration of computer and communications into the everyday affairs of mankind.

Kobayashi's vision was remarkably farsighted. When he gave this speech, computers were still locked up in air-conditioned rooms, guarded by a privileged few. I was one of them, the only person I knew with a 'personal' computer. The University of Chicago had purchased an IBM 360 Model 40 mainframe computer and installed it in the 'computer science'[1] laboratory, alongside the antique Maniac II computer, that a previous generation had built from discrete transistors, resistors, and hand-soldered wiring. They had installed the new computer in this shrine to already outmoded ways while they rewrote their software and figured out how to displace the IBM 7094 and its predecessor, the 7090, from 'the' computer room downstairs. The Physics Department arranged for me, a graduate student fresh from the farm, to use the new machine during the wee hours of the night to develop a program for doing molecular orbitals calculations in assembly language.

---

[1] I'll say much more later about why 'computer science' and 'software engineering' are usually quoted in this book. For now, it is sufficient to ask yourself the question, "Is there really any such thing ... yet?"

Fortran was regarded as 'too inefficient' for a calculation-intensive program that was sure to be heavily used!

Midnight shift, debugging assembly language code from a mainframe computer's console switches, was not what Kobayashi meant by integrating computers into the everyday affairs of mankind! However today, a quarter-century after Kobayashi's speech, computers are finally being integrated. In spades! First they targeted each company. Then each department. Then each desktop. Then each laptop. Now they're after our palmtops. Who knows where they'll stick them next? In our retinas? The Air Force is right on top of that. They've been building an experimental windowless aircraft in which every bit of information is computer-filtered and presented to the pilot through a technicolor, three-dimensional, WYSIWYG virtual reality with menu-driven object-oriented hypermedia GUI[2] and everything! Of course, they're not ready for the retinal implant stage quite yet. Their prototype is a headtop, a grotesque helmet that completely surrounds the pilot's head; so large that it is ludicrously impractical for high-G aerobatics. But with the present pace of miniaturization, who knows when they'll advance to Phase 2?

So much for the bleeding edge, the state of the art for integrating mankind with computers. For the leading edge, the state of present-day practice, consider the following ad from the Wall Street Journal as typifying what the writers for one computer manufacturer think Wall Street is eager to buy:

> *What happened to the old SPARC? Our maximum, 76 MIPS, is almost three times as fast as theirs. And our new workstation family starts at 57 MIPS—exactly twice Sun's maximum. The main reason for this enormous advance is our proven architecture. It enables our workstations to achieve a SPECmark of 72.2 versus their 21. These days, staying competitive is even more important than ever. Our workstations will give you the edge for as little as $12K for 57 MIPS and 17 MFLOPS. Or $20K for 76 MIPS and 22 MFLOPS. Call for more information. Then, instead of just striking a spark, you can set the world on fire.*
>
> — Full-page advertisement, *Wall Street Journal;* April 17, 1991

Just think! A full-page ad aimed right at the briefcase brigade. "Your days of anxious waiting are over! All you need to set the world on fire is a dash of MIPS and a dab of MFLOPS, seasoned with our proven architecture and its miraculous cure for the languid SPECmark." It makes me want to run right

---

[2]An outburst of common buzzwords. WYSIWYG means What You See is What You Get. GUI, or 'gooey', means Graphical User Interface.

out and buy one lest my clients discover that this ad was the first time I heard the term, SPECmark, used outside of janitorial services.

With all this talk of SPECmarks and MIPS, virtual realities and 32 bit color menu-driven GUIs, object-oriented this and hypermedia that, how did the ad-writers forget about the communications part of Kobayashi vision? While companies vie to provide MFLOPs, MIPs, and SPECmarks in bigger, better, cheaper bundles, customers are still expected to acquire computers, software, modems, and information services from a bewildering diversity of entirely different companies. Even today, with fast modems becoming commodity items and digital, even wireless, communications just over the horizon, the task of choosing, installing, and operating communication-based capabilities falls to users. Most of us are completely bewildered by the immense complexity of what should be the task of getting to the on-ramp of the information superhighway. Anyone who has experienced the despair and confusion of configuring Winsock or MacTCP, or figuring out what a "gateway address" is for, can appreciate the horrors that instructions like the following ones gloss over so glibly:

> *To use the recipes in this cookbook you will need a Macintosh computer with a direct connection to the Internet. You must have MacTCP installed on this computer. For full functionality, you must register an Internet name for this system with your Internet provider.*
>
> *If you are working in a school with no hope of a direct connection anytime soon, don't give up! Try to beg borrow or steal a connection from a local university or business with Internet. Once you set up the basic services you can put the server physically in their shop, and then maintain it and access it over the net! For this, you only need a SLIP, PPP, or ARA connection from your Internet provider.*
>
> *If you don't understand this page, sit down with your Internet provider and have a long talk. It won't help much to get help from the net; you really need to talk to your provider directly.*
>
> — Internet Server Kitchen for K12 Teachers,
> http://web66.coled.umn.edu/Cookbook/Kitchen.html

How can we expect ordinary mortals to assemble our own information age Model T Fords from the patchwork components we're being offered today? To buy a stand-alone computer for playing video games and running word processors or spreadsheets, customers need only visit Apple, or Compac, or Nintendo, and presto, the job is done. But to communicate with other people, a bewildering variety of choices must be considered and understood. Should I choose a broadband, or ethernet, or token-passing network? What software? What modem? What terminal emulator software? What internet access provider? What should I put in all those MacTCP or Winsock fields? And most of all, why doesn't the doggoned thing work?

And once these choices have been made, will I wind in touch with those I want to communicate with, particularly people like my technophobic Aunt Nellie home on the farm? Might I wind up in touch with those I'd rather *not* meet, such as hackers, crackers, and a ravening horde of marketeers and advertisers, eager to exploit the latest technological advances in pursuit of their interests even at the expense of my own?

1.2 INVISIBLE PLUMBING

A 21st century infrastructure would address many practical problems. For example, the government can serve as a catalyst for the private sector development of an advanced national communications network, which would help companies collaborate on research and design for advanced manufacturing; allow doctors across the country to access leading medical expertise; put immense educational resources at the fingertips of American teachers and students; open new avenues for disabled people to do things they can't do today; provide technical information to small businesses; and make telecommuting much easier. Such a network could do for the productivity of individuals at their places of work and learning what the interstate highway of the 1950s did for the productivity of the nation's travel and distribution system.
— Bill Clinton; September, 1992, *A Technology Policy for America:*
Six Broad Initiatives

A quarter-century after Kobayashi's keynote address, his vision is coming to pass. Computers are beginning to be thought of, not just as computational devices: 'personal', 'departmental', or 'desktop' computers, but as *communication* devices like televisions, telephones, and fax machines. And the President's enthusiasm is, of course, shared by the telecomputing industry, as can be seen in the following quotation from John Scully's speech to the newly elected administration, in the role he then held as Chief Executive Officer of Apple Computer:

We believe that the creation of a national information infrastructure must be a national priority, and we are willing to work in partnership with the government to see that it gets done. The development of an

*information infrastructure will raise the standard of living for all
Americans and enable our country to prosper in a competitive global
economy.*

The claim Scully made in this last sentence was clearly true of the canal,
railroad, and highway infrastructures of the manufacturing age. Even ordinary citizens, farmers on the prairies or gold prospectors in the mountains,
could immediately understand what a new railroad spur could mean to their
personal prosperity. But the miner and the farmer knew that their prosperity was derived from buying and selling the tangible goods of the manufacturing age, objects that are sufficiently hard to copy and to transport that
even ordinary people could derive revenue from providing them.

Similarly, we have no trouble understanding that building a nationwide
information infrastructure might do wonders for the prosperity of telecomputing companies like Apple. But the connection between infrastructure and
the prosperity of the *users* of that infrastructure is very tenuous indeed. Will
telecomputing networks "raise the standard of living for all Americans"? Or
the standard of living of telecomputing companies like Apple?

I often wonder what my technophobic Aunt Nellie back on the farm, or a
starving child in Somalia, will use a computer to do? Regardless of the glitz
and glamor that surrounds high-tech computing and communication technology, from the user's perspective, telecommunication infrastructures are
just so much plumbing to such people, a disagreeable subject with no interest unto itself. Ordinary people do care about content, about what the
plumbing might do for them, but they wish the plumbing would just vanish
altogether.

The explosive growth of the internet within the last few years (Fig. 1.2)
was largely due to the recent introduction of content-oriented tools such as
gopher and more recently the web. Web browsers are finally beginning to
make computers disappear, vanishing into the woodwork to no longer be a
topic of people's interest and concern, becoming just part of the plumbing,
doing its job invisibly and out of people's way. Computers are now just beginning to lose their nerdy reputation of being alienating **things**, ways to
avoid contact with other people. They're vanishing from sight, becoming
places, not *things*; transparent **windows**, communication channels through
which people communicate, cooperate, coordinate, and compete as members of a global electronically connected society.

Few would disagree with Kobayashi's vision today. However, we're not
there yet, and getting there involves fundamental changes in the way we
have thought of computers and telecommunications until now. Although the
term, paradigm shift, has been bludgeoned into uselessness by the trade

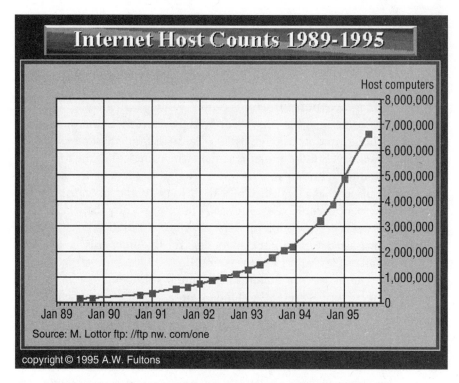

Figure 1.2 shows the explosive growth in Internet-connected computers since 1989. If this trend continues (and exponential trends never do), there could be a computer for each person on the planet by about the turn of the century....Presentation Slides by Tony Rutkowski; The Internet Society; ftp://ftp.isoc.org/isoc/charts

press, this is the best name for what we're facing. Like the Copernican Revolution, the Information Revolution involves a comparable shift in focus, away from today's techno-centric world that revolves around computer plumbing to a human-centric world centered on the people at the ends of the wires.

1.3 BREADTH VERSUS DEPTH

The DNA of Information: Bits and Atoms *The best way to appreciate the merits and consequences of being digital is to reflect on the difference between bits and atoms. While we are undoubtedly in an information age, most information is delivered to us in the form of*

atoms: newspapers, magazines, and books (like this one). Our economy
may be moving toward an information economy, but we measure
trade and we write our balance sheets with atoms in mind. GATT is
about atoms.

—Nicholas Negroponte, *Being Digital*

The explosive growth of computers and networks in the last decade has captured the attention of the media. However, the astounding growth rates in Fig. 1.2 tend to obscure the fact that there's also a half-empty side to this glass. For bits do have important deficiencies in relation to atoms. These deficiencies are in precisely the area that matters most to those who don't share technophiles' interest in telecom plumbing. By concentrating on breadth and ignoring the depth, we see the explosive spread of a glass of water spilled on the floor and imagine that we've discovered a new ocean.

I teach a course at George Mason University with the name, Taming the Electronic Frontier. A radical departure from traditional brick and mortar classrooms, this course doesn't teach. Rather it expedites *learning* by immersing students in first-hand experiences with the bewildering phenomena that are occurring all around the internet today. Learning occurs through experiential immersion, not by memorization of lectures. The class calendar presents tasks of steadily increasing difficulty with instructions on how to accomplish them successfully. The process begins with tasks that have students install Netscape and Eudora on their own computers at home, office, or laptop, and move to using these tools to research pages written by others, and ultimately to developing web pages for themselves. Students can also use preinstalled software on the GMU lab computers, but most buy a computer rather than drive to campus to do homework for this class.

The course does involve weekly lectures. These are carried locally by cable TV and microwave and remotely via T1 teleconferencing technology. For students for which these aren't an option, the lectures are on videotape in the library. The idea is to extend the idea of a "course" beyond academia's hallowed halls to working students who are increasingly unable and unwilling to fight traffic to comply with the industrial age's conception of what an educational experience should be.

Since the internet experience speaks so powerfully, I use the lectures to focus attention on the empty half of the glass, the electronic frontier issues that remain to be tamed. Yet in spite of exposure of a whole semester to these lectures, what invariably sticks in students' minds at exam time is the half-full part of the internet experience. They remember the part of the internet that was vital and strong and overlook the electronic frontier issues that remain to be tamed.

If growth rates continue, every company and individual might soon have a home page on the internet. This is clearly explosive growth, but it's primarily in breadth, not depth. When you examine the depth of information in the web pages out there already, this ocean is often shallow indeed. Internet hypesters regularly proclaim, "All the world's knowledge is on the internet. For FREE." With notable exceptions, as with any such generalization, the reality of the web is more like a stagnant lagoon, filled with advertising brochures, business cards, chitchat, and freeware. This is entirely unlike the thriving oceanic ecology of any good library, bookstore, or newsstand.

To bring the half-empty aspect of the glass into focus, I've recently added a task that has students measure the depth of the web at any point in its breadth. Try this yourself. Choose any company with a presence on the web. Measure the depth at that point by simply counting the bytes. Include whatever local links may be there but exclude external links not on that server. Then contrast this measurement with a back-of-the-envelope estimate of the information in the real company. For a publishing company like Addison-Wesley, the estimate should include the books they sell as their products, the information in their filing cabinets, notebooks, computers, and the knowledge in their employees' heads. Such comparisons will show that the depth of information in any real company is many times greater than the depth of information in its web.

Although the breadth of the web clearly has great appeal to those of a technical persuasion, to the ordinary people who buy published products, this shallowness is disappointing indeed. To most people, breadth isn't a virtue but a problem, something to be overcome by searching or browsing. People don't pay to acquire new problems. They pay to acquire *depth*, which is precisely what the internet is most lacking today.

1.4 THE ELECTRONIC FRONTIER

> *Like distant islands sundered by the sea,*
> *We had no sense of one community.*
> *We lived and worked apart and rarely knew*
> *that others searched with us for knowledge, too.*

> *Distant ARPA spurred us in our quest*
> *and for our part we worked and put to test*
> *new thoughts and theories of computing art;=*
> *we deemed it science not, but made a start*

Each time a new machine was built and sold,
we'd add it to our list of needs and told
our source of funds "Alas! Our knowledge loom
will halt 'til it's in our computer room.

But, could these new resources not be shared?
Let links be built; machines and men be paired!
Let distance be no barrier! They set
that goal: design and build the ARPANET!

—Vint Cerf, Internet pioneer

Barlow's electronic frontier image evokes a vision of limitless opportunities of a world without bounds. But it also reminds us that frontiers are places where established values break down and new ones get erected, where established communities get pushed aside and new ones move in. The collision between American Indian and European cultures, culminating in the former's virtual extermination during what the latter chose to call the "taming" of the wild west is suggestive of how the taming of the electronic frontier may yet unfold.

At the heart of this tragedy were two opposing conceptions of property. Although indigenous American Indian cultures did allow individuals to own property, this never had anything close to the priority that ownership held in European cultures. Some tribes even held regular Christmas-like celebrations called potlatches, during which individuals gave away much of their property to others. There was nothing comparable to the European's notion of 'real estate', land that individuals or even communities could buy, sell, or own. From the viewpoint of the European newcomers, this vast continent was unowned and ripe for the taking. So take it they did.

The electronic frontier also has an indigenous property-adverse culture. This culture still has the upper hand today, but is just beginning its encounter with an invading horde of property-conscious colonizers. The newcomers are not organized yet, and don't have very clear ideas of what this electronic frontier is *for*. The newcomers are still wandering around, wondering what all the excitement is about. But the air is electric. Anything can happen once somebody figures out an answer to the newcomers' biggest question, "How the heck can honest folk make a living out here?", a question that has property-consciousness right at its heart.

The natives are those who originated the tangled conglomeration of computers, software, databases, and telecommunication links that collectively constitute the Internet today. Part of its history dates back to the Department of Defense's interest in telecommunication technology that might survive failure of any component. They provided funding to various universities and companies to develop the packet-switching technology upon which much of the Internet relies: the ARPAnet. Other parts derived from the anarchic collaboration of students and programmers who simultaneously built a low-tech network (called Usenet) based on dial-up modem technology (circuit switch) capable of carrying messages (called Netnews) more or less reliably between one machine and another.

1.5 THE LARGEST UPHEAVAL OF ALL

U.S. OFFERS $2 MILLION, RELOCATION FOR TIPS ON WHEREABOUTS OF PABLO ESCOBAR WASHINGTON (AUG. 15 1992) UPI - U.S. officials Friday offered a reward of up to $2 million for information leading to the arrest of escaped Colombian drug lord Pablo Escobar. To date, officials have paid out some $2.3 million in such rewards—$2 million for anti-terror tips, and $350,000 for narcotics-related information.
— United Press International, from America Online

A big part of the transition to information-age commerce involves figuring out how organizations can meet payroll as an ever-increasing percentage of their workforce is engaged in intangible white-collar work. Although packaging their work as electronic goods is only part of the answer, this often involves figuring out how to buy, to sell, and to own intangible electronic objects. For these are a strange kind of 'goods' that can be copied so easily that they propagate at the speed of light through the information age infrastructures that are so much a part of today's increasingly global economy.

The telecomputing industry has been driven until now by two distinct constituencies. The first is the minuscule minority of the human race who are interested in media for artistic expression. This community pursues the vision that Alan Kay's group at Xerox Parc enunciated during the 1970s as the 'Dynabook', a notebook-sized computer that would allow us to create with our personal computer under a tree in the orchard, working in isolation from everything around us. The personal computing revolution largely targeted the needs of this minority who thinks of computers as a medium of artistic expression. Most successful computer applications are still targeted at creative folks who see a blank word processor or spreadsheet page as an irresistible incentive, electronic canvases on which to create.

The second constituency is more numerous, but still a very small fraction of the population at large. These are those centrally planned bureaucracies that constitute the 'business market'. This is the ever-increasing part of the workforce, the white collar workers producing information for others in the organization to use. The desire to 'do something' about white-collar productivity, and to get away from the power struggles of old-style Management Information Systems establishment, was the second of the forces that supported the personal computer revolution.

The third constituency is not yet upon us, but it's by far the largest market of all. These are the information consumers like my stereotypical Aunt Nellie back on the farm. To this segment of our community, blank canvases are hardly an incentive to creative inspiration. Their blank emptiness is a terrifying obstacle, completely irrelevant to the interests of this market. Trying to sell Aunt Nellie blank electronic canvases like word processors and spreadsheets is like peddling rolls of blank newspaper and printing presses to the masses in the hope that they would rather compose their own news.

The largest market of all is the one that the computer industry has never managed to reach. The majority want their canvases already painted, as ready-to-use information-age goods to be bought, learned from, and enjoyed right off the shelf. They are completely indifferent to telecomputing infrastructure. They do care about the water but have no interest in the plumbing, which is what the computer industry is offering them today. They want well-written relevant information-age goods such as timely online articles of events around the world.

The unsatisfied interests of this vast majority of ordinary people, bemused and bewildered by the PC revolution pioneers' interest in blank electronic canvases, are the spark that could ignite the largest upheaval of all. This is the creation, for the first time in human history, of a true information economy in which intangible information-age goods can be bought and sold as robustly as tangible manufacturing-age goods are bought and sold today.

1.6 TAMING THE ELECTRONIC FRONTIER

We do not really understand how to live in cyberspace yet. We are feeling our way into it, blundering about. That is not surprising. Our lives in the physical world, the "real" world, are also far from perfect, despite a lot more practice. Human lives, real lives, are imperfect by their nature, and there are human beings in cyberspace. The way we live in cyberspace is a funhouse mirror of the way we live in the real world. We take both our advantages and our troubles with us.

— Bruce Sterling, *The Hacker Crackdown*

On Sunday evenings, the local bluegrass radio station broadcasts audio reruns of *Gunsmoke*, the TV series that was popular in my youth. Marshal Dillon, Chester, Doc, and Lilly provide nostalgic glimpses of the long-gone mythos of the American West, so remote from today's sanitized, politically correct era. Even violent death was not cause for Marshal Dillon's concern so long as the victim was armed, shot from the front, and other proprieties of this era were observed.

Although virtual gunslinging is as rife today on mailing lists and newsgroups as six-gun dueling was in the wild west, human life is no longer at stake. Today's problem is different but equally destructive of social order. Human life is safe, but property is entirely at risk. In fact, property is so thoroughly at risk that the very possibility of robust ownership of electronic property is in dispute. This dispute is already in the news today, but with names like privacy and piracy and in connection with technologies like the Clipper Chip. The name electronic property isn't generally used in these disputes, but the absence of robust production for electronic property rights is the common thread all the same.

The gold bullion on the Wells Fargo stagecoach, the rancher's cattle, and the deed Snively Whiplash tried to get from Little Nellie were tangible goods made of atoms. Goods made of atoms abide by the natural laws that have underpinned commerce, and skullduggery, since antiquity. Physical conservation laws, such as the law of conservation of mass, support and enforce the human meanings of such terms as to buy, sell, own, or steal.

But quite recently, only within the last generation, mankind has experienced a new form of goods that is immune to the conservation laws that have supported commerce since antiquity. Electronic property is made of bits. These don't abide by the physical laws that apply to goods made of atoms. Bits can be replicated in nanoseconds and transported at literally the speed of light. Although this is largely responsible for the growing importance of such goods, it also leads to a breakdown of the understandings that supported collaborative enterprise in the past. Indeed, establishing a robust meaning of the term, electronic property, is the pivotal issue that must be resolved before the electronic frontier can ever be tamed.

I'll use this term, electronic frontier, to include subtopics, such as software engineering and computer science, that are typically excluded from this frame. The term was popularized by John Perry Barlow and Mitch Kapor of the Electronic Frontier Foundation, where it applies primarily to issues arising in connection with network-based computing and communication. In the worldview and terminology I'll adopt in this book, this is only the topmost level of an extremely broad and deep tree of issues. Although these issues are rarely discussed within the same frame, they all share, and are related by, the same pivotal issue mentioned above. They all involve what

I'll call electronic property in this book. These are goods made of bits, as distinct from the physical atoms from which goods have been composed since antiquity.

As might be imagined from the claim of being the pivotal issue for information-age enterprise, this tree of issues is both broad and deep. Its breadth encompasses issues that are generally encompassed by the electronic frontier headline—everything from electronic mail, to FTP archives of computer software, to web pages—and the issues of privacy and piracy that such topics readily evoke.

It is less common, and far more difficult, to bring the depth of this tree into the frame. The electronic frontier community is articulate about ownership issues of network-based objects such as E-mail and web pages. However, there is very little recognition of the fact that the very same issues apply at every level of architectural granularity, right down to the smallest components from which large-granularity objects are composed. For example, the same issues of privacy and piracy that arise in connection with web pages arise with even greater severity with respect to clip art, quotations, and document links used in constructing that page. These same issues arise not just with web pages, but with the computer software (web browsers) used to build and display the pages. And the same issues arise with even greater severity with respect to the small-granularity reusable software components that are increasingly being sought after as a way of reducing crippling software development costs.

Therefore this book will use "electronic frontier" as an umbrella term. It encompasses information-age enterprise in its full breadth and depth. In particular, it includes software engineering and computer science as interesting special cases within the same umbrella, not as special fields shielded by their obscurity from consideration by those who don't specialize in these fields.

Computer software, reusable software components, and multimedia documents are not pumped from wells, dug from mines, or harvested from sun-drenched fields. Unlike petroleum, silicon, or food products, which are ultimately composed of tangible atoms derived from nature, electronic property is not derived from nature but entirely from the activities of man.

The age of electronic reproduction has thrust mankind into a new era in which more and more of us are building goods that consist entirely of capital, knowledge, and labor, and contain nothing whatsoever that abides by the fundamental conservation laws of nature. Electronic property is trivially replicable in a way that gold bullion in the stagecoach strongbox and the rancher's land and cattle never were. Can there be any notion of buying, selling, or owning in a medium within which "goods" can be replicated in nanoseconds and transported at literally the speed of light? And so long as this disassociation from nature undercuts fundamental human understandings such as ownership, where will humans receive the incentive to invest their labor, capital, and knowledge in producing high-quality electronic goods?

This poses the paradox that will concern us throughout this book: "Now that all the computer and telecommunication industries are putting ever better, faster, less expensive telecomputing technologies at everyone's disposal, who will produce content for this carrier to convey?" How can people, not just big corporations but ordinary citizens, earn their living in a medium in which their ownership iş entirely at risk? So long as there's no robust way to earn an honest living on this new frontier, wouldn't we be better off staying put in the Industrial Age? Although industrial-age jobs like flipping burgers don't pay well, they do pay. This is more than can be said for goods that can be copied in nanoseconds and transported at the speed of light.

Koji Kobayashi's prediction of the integration of computers, communications, and mankind is well on its way to becoming a reality. Everyone agrees that the telecomputing industry will continue to deliver ever better, faster, and less expensive information-age plumbing through which information-age goods will eventually flow, like water from pump to tap, or more precisely, from information providers to consumers, since electronic goods originate solely from human labor.

The techno-centric worldview of those fields has made great progress in the generation it has been in existence. Object technology is only the most recent of many examples. Nonetheless, software engineering still seems unable to achieve the rapid improvement in software reliability, cost, quantity, or quality that hardware engineering delivers with clockwork regularity. Regardless of techno-centric innovations, the software industry remains firmly in the grip of the software crisis, even in the midst of the very period when the hardware industry has become famous for exponentially improving productivity.

Nor is the problem confined to computer software. The same problem recurs at every level of granularity, including the highest level common today, hypertext documents on the worldwide web. At first glance, the phenomenal growth of the web, nearly 40% per month the last time I checked, overwhelms us. With hypertext carrier capability growing so rapidly, and with so much content already available, it is easy to forget that the information we are most interested in is rarely available there. While the web is sufficiently complex to defy any simple characterization, the web is more like a huge cross-linked collection of business cards and brochures than a library full of books representing the world's collective wisdom. The kinds of information that are available in such abundance in industrial age institutions like libraries, bookstores, and newsstands aren't likely to ever be electronically available until endemic ownership and revenue collection issues have been resolved.

Magazines and books are tangible items that people can earn money producing. Business cards and brochures are given away free. Tangible goods are made of atoms and this makes it completely clear what it means to buy, sell, and own them. Every urchin in the souk can understand exactly how to engage in commerce with tangible goods. But computer software and web pages are intangible goods made of bits. It is entirely ambiguous what it means to 'own' goods that can be copied and transported in nanoseconds.

1.7 INDIGENOUS ENCOUNTERS ON THE ELECTRONIC FRONTIER

Initially, to seasoned members of the IT community, the home video game industry must have seemed frivolous and irrelevant. Designed for children and marketed as toys, there was little about the early models of the home video game to inspire interest or concern in the computer and communications industries. Yet today, in many ways, the HVG industry has already become a more strategic, more lucrative, and more international business than many thought possible.
— Home Video Game Market Overview. JULY 1995,
by Chris Stiles, Technology Analyst, ATIP Tokyo

The users outfoxed us again. It happens every fifteen years or so in this business. We lost our grounding, the users rebelled, and a new incarnation of the software business has been created. What is it? The Internet, of course. It's a very magic thing whose potential has barely been explored. New stuff is happening almost on a daily basis. There's a rebellious spirit to it. What are they rebelling against? The greedy VCs that funded the software industry who were too busy inventing scams to make Mitch Kapor-style money. (The scams didn't work.) And they're rebelling against Bill Gates, who has already made Mitch Kapor-style money, many times over, and possesses something much more offensive—the power of FUD: Fear, Uncertainty and Doubt.
— Amusing Rants from Dave Winer's Desktop

In school we learn that salt is composed of sodium and chloride ions bound together to form a crystal by powerful coulombic forces originating from the opposing charges of these ions. But we also know that if we drop such a crystal it will fall to the floor. The shock of the fall might break it, thus disrupting ionic bonds no matter how powerful. Although the ionic forces are more powerful, they operate only at short (atomic scale) distances. Over much longer distances of everyday experience, the force of gravity plays an even more significant role. I find this analogy helpful when thinking about the controversies that inevitably arise, in academic circles at least, when money is mentioned in other than a negative light. Left vs Right, Democrat vs Republican, Academia vs Industry, Communitarianism vs Capitalism. It's as if the short- and long-range forces that coexist in something as simple as a crystal of salt can't coexist just as peacefully in the overwhelming complexity of human society.

Partially because my wife, Etta, is an American Indian (Choctaw), we've decorated our home in Southwestern style. We've collected Indian rugs, baskets, and pottery, supplementing them with African and Australian native arts whose beauty and sophistication we find deeply comforting in an unsettling high-tech era. We occasionally wonder why modern industrial technology seems incapable of producing such treasures, and why the indigenous cultures that produced them are in retreat from modern society, not just in America but around the globe. And why the communitarian ideal that undergirds such societies can't coexist with the capitalistic ideals of ownership, much as short- and long-range forces coexist within the physical world.

I'm one of the electronic frontier's indigenous people, a member of the Nerd tribe. I spent my career in the techno-centric worldview characteristic of software engineering and computer science. And I'm trying to figure out what the recent influx of property-conscious settlers, the Newbies, will mean in the end. Taming the West involved displacing the established worldview of the natives with the foreign worldview of European newcomers. The worldview of the pioneers of the electronic frontier is starting to collide with those who may displace us someday, ordinary folks who are uninterested in exploring but are interested in how to earn their living there.

Internet history revolved around the communitarian ideal that structured the native cultures of America, Africa, and Australia. By this I mean the short-range social forces that govern behavior within family and tribe, as distinct from the long-range forces arising from commerce. These longer-range forces work across vast distances in both space and time, helping people who have never met (and might not get along well if they did) to cooperate in vast ventures that span the entire globe. Although the communitarian ideal of sharing has a strong ethical appeal, it is also clear that communitarian cultures can rarely compete effectively against cultures structured around the longer-range forces of commercial exchange transactions.

The transition from exploration-minded natives to property-conscious newcomers is only just beginning on the electronic frontier. Although I'll survey some of these commerce-oriented initiatives later, my goal is not to provide yet another description of the electronic frontier as we find it today. The objective is to anticipate, and perhaps even to guide, this evolution in the hope of avoiding the tragedies of indigenous/newcomer encounters of the past.

That, of course, is the bad news. The good news, from the point of the indigenous Nerd tribe at least, is that commercial exchange transactions work effectively only for tangible goods made of atoms. They're ineffective for the intangible goods of the electronic frontier. Since such goods can be replicated and transported instantaneously, there is no robust basis for established human-centric understandings such as what it means to buy, sell, and own.

The paradox of the electronic frontier is that in spite of its vast potential, we have never figured out what it means, or what it *should* mean, to buy, sell, and own goods that can be copied and transported so readily. But mankind will resolve this paradox because we have no other choice. Turning back the clock to the age of manufacturing when things were so much simpler is no longer an option. The information age is with us to stay.

Chapter 2

Structure of Production

The new international airport in Denver was to be the pride of the Rockies, a wonder of modern engineering. Twice the size of Manhattan, 10 times the breadth of Heathrow, the airport is big enough to land three jets simultaneously—in bad weather. Even more impressive than its girth is the subterranean baggage-handling system of the airport. Tearing like intelligent coal-mine cars along 21 miles of steel track, 4000 independent "telecars" route and deliver luggage among the counters, gates, and claim areas of 20 different airlines. A central nervous system of some 100 computers networked to one another and to 5000 electric eyes, 400 radio receivers, and 56 bar-code scanners orchestrates the safe and timely arrival of every valise and ski bag.

At least that is the plan. For nine months, this Gulliver has been held captive by Lilliputians—errors in the software that controls its automated baggage system. Scheduled for takeoff by last Halloween, the grand opening of the airport was postponed until December to allow BAE Automated Systems time to flush the gremlins out of its $193-million system. December yielded to March. March slipped to May. In June the planners of the airport, their bond rating demoted to junk and their budget hemorrhaging red ink at the rate of $1.1 million a day in interest and operating costs, conceded that they could not predict when the baggage system would stabilize enough for the airport to open.

— "Software's Chronic Crisis; Trends In Computing", by W. Wayt Gibbs, *Scientific American*; September 1994; Page 86

Like fish in the sea and birds in the air, we are immersed in the social order that arose from the industrial revolution. We live in houses and apartments that weren't even possible before the industrial revolution created the components that made them possible. We live in suburbs that could never have evolved without the highways, subways, and automobiles that emerged from the industrial revolution. And we work in organizations that evolved to thrive in this industrial-age milieu, exactly as fish evolved to thrive in the sea.

This milieu is so pervasive that it is easy to forget that we are all embedded in it. Seeing it involves bringing a commonplace into view, something that is concealed by its obviousness. The formal name of the commonplace we'll be concerned with in this chapter is *structure of production*. This is the human-centric system that hides the complexity of tangible goods, doing it so thoroughly that we can easily forget that even simple manufactured objects can be massively complex inside. Since structure of production barely exists for electronic goods, except in the most rudimentary form, we experience breakdowns, whose general name is the software crisis, such as the Denver International Airport debacle.

The Denver International Airport debacle brings two very different social orders from opposite ends of time's arrow into juxtaposition. Both of these social orders have been successful at one time or another, but they almost invariably clash whenever they meet on the historical frontiers around the globe:

- The **industrial** social order that successfully produced the tangible goods of this example: the baggage handling carts, runways, and airplanes and the vast number of raw materials and intermediate-level subcomponents from which the baggage carts, runways, and airplanes were constructed.

- The **artistic** social order that failed to deliver the electronic goods of this example on time and within budget. These goods include not only the top-level computer software, but the smaller components from which this software might have been composed had they been produced by a comparably advanced social order.

This chapter will juxtapose several industrial and artistic products in order to bring the diverse social orders that produced them into focus. The purpose of this comparison is not to argue that any one of these goods is more or less complicated than the others, and certainly not to argue that one or the other of the social orders is "better" in absolute terms. The purpose is to reveal differences between social orders that would otherwise be hidden by their obviousness. The purpose is to show why the social order we use today to build electronic products has been unable to master complexity. The problem is not primarily a failure in our tools, methodologies, or programming language. The failure originates in the fact that, in spite of their many virtues, artistic social orders differ from industrial orders in not mastering the complex management techniques that make us think of industrial orders as mature.

2.1 THE TALE OF THE WOODEN PENCIL

The Denver International Airport involves a vast variety of products, any of which would have served for this demonstration. I might have chosen the airplanes, buildings, highways, or runways because everyone knows that these are complex. I could have shown how the food in the restaurants was grown, harvested, processed, distributed, marketed, and sold by the vast agricultural networks of the post-agrarian revolution era. I could have chosen the medical products in the infirmary, the cleaning products for the janitors, the fuel for the planes, the ammunition for the police, or the computers and telecommunication systems at the reservation desk.

But since anything would suffice, I picked what we might perceive as the simplest, the ordinary wooden pencils of Fig. 2.1 that ticket agents might use for their paperwork.

Figure 2.1 The simplest of objects, such as the ordinary wooden pencils in this cup, can teach profound lessons as to how complexity can be encapsulated so that we even forget that it's there.

Programmers think of pencils as mundane and simple, certainly in comparison with electronic equivalents, word processors. However, the following tale of this mundane object unveils the vast complexity that is hidden inside manufactured objects, hidden so completely that we forget that something has hidden it from view. The tale of the wooden pencil exposes this system so that software engineering might someday notice and adopt it to master the overwhelming complexity of computer software:

I, Pencil, simple though I appear to be, merit your wonder and awe, a claim I shall attempt to prove. In fact, if you can understand me—no, that's too much to ask of anyone—if you can become aware of the miraculousness that I symbolize, you can help save the freedom mankind is so unhappily losing. I have a profound lesson to teach. And I can teach this lesson better than can an automobile or an airplane or a mechanical dishwasher because—well, because I am seemingly so simple.

Simple? Yet, not a single person on the face of this earth knows how to make me. This sounds fantastic, doesn't it? Especially when it is realized that there are about one and one-half billion of my kind produced in the U.S.A. each year.

Pick me up and look me over. What do you see? Not much meets the eye—there's some wood, lacquer, the printed labeling, graphite carbon, a bit of metal, and an eraser. Just as you cannot trace your family tree back very far, so is it impossible for me to name and explain all my antecedents. But I would like to suggest enough of them to impress upon you the richness and complexity of my background.

My family tree begins with what in fact is a tree, a cedar of straight grain that grows in Northern California and Oregon. Now contemplate all the saws and trucks and rope and the countless other gear used in harvesting and carting the cedar logs to the railroad siding. Think of all the persons and the numberless skills that went into their fabrication: the mining of ore, the making of steel and its refinement into saws, axes, motors; the growing of hemp and bringing it through all the states to heavy and strong rope; the logging camps with their beds and mess halls, the cookery and the raising of all the foods. Why, untold thousands of persons had a hand in every cup of coffee the loggers drink!

The logs are shipped to a mill in San Leandro, California. Can you imagine the individuals who make flat cars and rails and railroad engines and who construct and install the communication systems incidental thereto? These legions are among my antecedents.

Consider the millwork in San Leandro. The cedar logs are cut into small, pencil-length slats less than one-fourth of an inch in thickness. These are kiln dried and then tinted for the same reason women put rouge on their faces. People prefer that I look pretty, not a pallid white. The slats are waxed and kiln dried again. How many skills went into the making of the tint and the kilns, into supplying the heat, the light and power, the belts, motors, and all the other things a mill requires? Sweepers in the mill among my ancestors? Yes, and included are the men who poured the concrete for the dam of a Pacific Gas & Electric Company hydro plant, which supplies the mill's power.

My "lead" itself—it contains no lead at all—is complex. The graphite is mined in Sri Lanka. Consider these miners and those who make their many tools and the makers of the paper sacks in which the graphite is shipped and those who make the string that ties the sacks and those who put them aboard ships and those who make the ships. Even the lighthouse keepers along the way assisted in my birth—and the harbor pilots.

The graphite is mixed with clay from Mississippi in which ammonium hydroxide is used in the refining process. Then wetting agents are added such as sulfonated tallow, animal fats chemically reacted with sulfuric acid. After passing through numerous machines, the mixture finally appears as endless extrusions, as from a sausage grinder, cut to size, dried, and baked for several hours at 1850 degrees Fahrenheit. To increase their strength and smoothness the "leads" are then treated with a hot mixture that includes candelilla wax from Mexico, paraffin wax, and hydrogenated natural fats.

My cedar receives six coats of lacquer. Do you know all of the ingredients of lacquer? Who would think that the growers of castor beans and the refiners of castor oil are a part of it? They are. Why,

even the processes by which the lacquer is made a beautiful yellow
involves the skills of more persons than one can enumerate!
— I, Pencil, my family tree as told to Leonard E. Read[1]

The tale of the pencil shows a vast diversity of commercial enterprises that manage to cooperate effectively within a vast social order. People who probably never met, and might not get along if they did, managed to cooperate so efficiently that we forget that they needed to cooperate at all. The lumberjack in the cedar forest and the rape seed oil farmer in the Dutch East Indies are intertwined with the ever-changing needs and tastes of airport workers in Denver, the school children in Kansas, and the office workers in New York, with everyone who purchases their products.

Yet the slightest changes in fashion and demand are efficiently communicated by price signaling across the geographic, language, and cultural barriers that might easily prevent cooperation in any other manner.

2.2 STRUCTURE OF PRODUCTION

The peculiar character of the problem of a rational economic order is
determined precisely by the fact that the knowledge of the
circumstances of which we must make use never exists in concentrated
or integrated form but solely as the dispersed bits of incomplete and
frequently contradictory knowledge that all the separate individuals
possess. The economic problem of society is thus not merely a problem
of how to allocate "given" resources—if "given" is taken to mean
given to a single mind, which deliberately solves the problem set by
these "data." It is rather a problem of how to secure the best use of
resources known to any of the members of society, for ends whose
relative importance only these individuals know. Or, to put it briefly, it
is a problem of the utilization of knowledge that is not given to anyone
in its totality.
— "The Use of Knowledge in Society", Friedrich Hayek, *American Economic Review,*
XXXV, No. 4; September, 1945, 519-530.

This quotation comes from the leading figure of a body of work called *Austrian Economics*. This branch of economics rejects the social fashions that became popular during World War II. It started coming back into favor when

---

[1] I acquired the paper from which I extracted these quotes years ago as a dog-eared Xerox copy that held no trace of its publisher or author's address. I was so impressed with it that I OCR'd it into electronic form where it became a mainstay of my courses on electronic property. I have since learned that it was published in *The Foreman*, 1956, Foundation for Economic Education; Irvington-On-Hudson, NY 10533. Used by permission.

the spectacular collapse of the Russian Empire confirmed Hayek's notion of price signaling as a mechanism whereby all participants in a system can mobilize their tacit knowledge of time, space, scarcity, abundance, preferences, and taste. Austrian Economics rejects the mainstream economics infatuation with mathematical models that can only be solved once economic systems are "at equilibrium." The Austrians point out, quite rightly in my opinion, that living systems come into equilibrium with their environment only when they die.

Structure of production is a foundation concept of this book. It means a network of human understandings that binds disparate actors into an organic and dynamic economic system. The pencil story shows how a structure of production achieves cooperation across vast distances in space and time, producing the manufactured goods that surround us in such profusion today.

Structure of production delivered the tangible goods for the Denver International Airport on time and within budget. Therefore it seems worthwhile for the software engineering culture, the culture that failed there, to look very closely at how its neighboring culture accomplished its success in considerable detail. Again, this is not because these details are more than just common sense, but that their obviousness makes them so easy to overlook.

Figure 2.2 eliminates the details of the pencil story to highlight only the bare skeleton of this system. The circles in this figure represent independent economic agents. These are often individual people such as the individual rape seed farmer or lumberjack. But they can be, and often are, the organized collections of individuals we call firms, such as the mill in San Leandro. Since the differences between individuals and firms, and the precise nature of their products, aren't material here, I've backgrounded the difference by representing everything as homogeneous circles.

The lines represent long-range bonding forces, the web of commercial commitments that joins what would otherwise be independent agents into a cohesive social order. Miraculously this web accomplishes this coordination entirely without coercion. It yields a system capable of producing as many pencils as we might need, from independent actors that are actively eager to cooperate with each other in responding to the slightest whim people they've never met and might very well not get along with if they had.

That is really a pretty amazing "machine" in that figure, to use the kind of industrial-age metaphor that I'll criticize in a moment. It is as if those lines are pipes in a vast hydraulic system that connects the graphite tip of each and every pencil to the farthest reaches of the globe. These pipes transmit the consumption of each and every pencil, the faintest trace of graphite on paper, to every stage of the process for producing more regardless of where they might be on this planet.

Customers

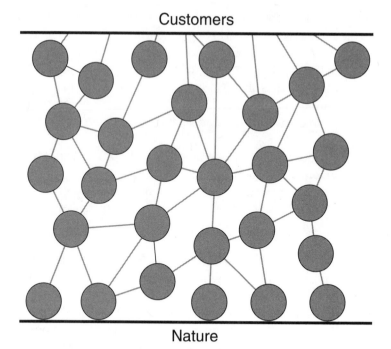

Nature

Figure 2.2   strips all of the details from the story of how pencils are provided to highlight the key feature of the social order that provides them. Individuals and institutions are bound into a cohesive social order by the long-range binding forces that were described in the *I, Pencil* example.

Each trace on paper or ground away in the pencil sharpener creates a vacuum that moves without loss throughout this system. This pulse automatically, nearly magically, draws *precisely the right amount* of production from **every stage** in this vast system.[2] The pulse travels to the rape seed oil farmer,

---

[2] This view of markets is an outcome of a branch of economics called the *Austrian* School because of the nationality of its most prominent founders, Friedrich Hayek and Ludwig Mises. For those who would like to read further, I particularly recommend *The Use of Knowledge in Society*, the paper for which Dr. Hayek won the Nobel Peace Price. It is an entirely approachable way of entering a fascinating but massive literature that originated from the "calculation debate" of the 1930s. This debate originated from a formal challenge that the Austrian school issued to the socialist central planners of that era. The challenge was to explain how collectivist central planning bureaus could ever acquire enough distributed tacit knowledge to match the knowledge gathering and processing ability of an entire population of self-interested individuals coordinated through markets. Although the Austrian school fell into decline when Keynesian Economics came into fashion during World War II, stubborn defenders persisted nonetheless. The founders of my own department at George Mason University, Don Lavoie and Jack High, previously of the GMU Economics Department, were actively involved in this struggle. The recent collapse of the centrally planned Russian economy is now regarded as confirmation of the Austrian position on the calculation debate, and it is finally receiving the attention it deserves.

the cadmium sulfide miner, the lumberjack, and each higher level that uses their products, and persuades them to increase or decrease production to **exactly the right level** that the right number of pencils will be produced and not one pencil more. As demand for pencils and pencil subcomponents increases or decreases, so varies the production, right down to the most minuscule of components.

2.3 SOCIAL BINDING FORCES

Given our propensity for misunderstanding those even marginally different from ourselves, isn't any mechanism that can get thoroughly different people to cooperate this effectively worth very close scrutiny indeed?

Since this market mechanism binds individuals and organisms into a cohesive social order, I'll call it a binding force in this book. Force has exactly the same meaning as in physics, something that can achieve action-at-a-distance. To avoid confusion with other binding forces to be introduced later, I'll refer to the commercial cooperation emphasized in this story as a **long-range binding force**, a force that operates at longer ranges than those that govern social interaction within face-to-face communities such as families, companies, and tribes.

Physics makes the same distinction between short- and long-range forces for the binding forces of matter. Four of these binding forces were known when I studied physics in school, and I've heard that a fifth has since been discovered. Isaac Newton is generally credited with discovering the first and weakest of these forces, gravity. This is the longest range of the forces for it reaches out indefinitely to govern the paths of planets and stars in the heavens.

However, gravity is also by far the weakest of these forces. It is negligibly weak in comparison with the shorter-range electrostatic forces that govern the chemical reactions that produced the planets, rocks, homes, and furniture of everyday life. These chemical forces are in turn far weaker than the awesome nuclear power that originates from the extremely short-range forces that bind subnuclear particles into atoms.

I chose this term, *long-range forces*, in response to those who believe that commerce is not, or should not be, the deciding factor in human affairs. This argument is especially common on the electronic frontier at the moment, where so many of its most significant advances have been achieved through initiatives that were anything but commercial.

Long range does not imply that this force is more powerful than the shorter-range forces to be discussed later. Short-range forces are far more powerful, dominating our everyday actions so thoroughly that the longer-range forces can too easily be forgotten. For example, communitarian forces such as concern for the well-being of our family and the surge of goodwill and energy from a pat on the back by our employer usually exert far more influence than the longer-range forces to be emphasized here.

2.4 COMMERCIAL EXCHANGE TRANSACTIONS

Just as the forces of physics can be expressed as interactions mediated by photons, the long-range force that binds individuals into cooperating members of a pencil-producing social order originates from a comparable quantum event. The name economists use for this quantum event is the *commercial exchange transaction*.

What makes this event so powerful is its utter simplicity and ubiquity. No formal education in the laws of physics is needed to know that regardless of how a vendor's product is consumed, it will renew the vendor's opportunity to sell more. The West Indies rape seed farmer can rationally decide whether to continue playing his role in the pencil production system, year after year, in competition with other ways he might invest his time, labor, and land. If he chooses to continue in, or defect from, this game, it is on the basis of the *price* he can charge for whatever goods he might produce.

The miniscule pulse of hydraulic energy generated by the consumption of a pencil in Denver is conveyed to the farmer as a correspondingly minute increase in price. The mechanism that conveys this pulse and ensures that it isn't lost in the noise is none other than conservation of mass.

This concept of *price* is the result of a social, market-based process in which buyers and sellers leverage their tacit knowledge of scarcity and demand to their mutual advantage. I will have little to say about prices in this book for prices are driven by this social process and are independent of anything I might say or do. Rather, I will focus on conservation of mass as that which supports the commercial exchange transaction, the quantum event upon which the whole system relies for its energy and upon which the notion of price is based.

The meaning of commercial *exchange* transaction involves the knowledge that tangible goods are *scarce*. The rape seed oil farmer knows that each and every sale of a pencil in Denver will ultimately, and inevitably, replenish the market for rape seed oil thousands of miles away.

It is crucial to notice that consumption of this pencil doesn't affect only the prosperity of any particular level of this hierarchy. In particular, it doesn't affect only the office supply store at the very top of the food chain. As the stock of pencils in the store is depleted, a pulse of scarcity is produced that will be felt by *every* component in the tree. The money that the office supply store collects will soon be spent to buy more, generating a pulse of prosperity that travels right to the bottom of the hierarchy, to the lumberjacks, rape seed farmers, and mill operators throughout the global structure of production.

All this is nearly independent of geographic proximity. So it amounts to a long-range binding force, something that can bind self-interested individuals into cohesive and cooperative societies. The rape seed farmer in the West Indies doesn't need to know, like, or even trust the office supply store owner in Denver, to devote his life to serving their needs and whims. The vacuum that links the farmer, the office supply operator, and the pencil-wielding clerk doesn't even depend very greatly on human understanding or even human law. The vacuum is not maintained by the law of mankind. The system runs on the natural laws that govern all goods made of atoms, laws such as conservation of mass.

2.5 THE LAIR OF THE SOFTWARE WEREWOLF

This image of a vast hydraulic system filled with an incompressible fluid helps to see exactly what's wrong with software. Tangible goods are governed by physical law, conservation of mass. This provides the fluid that's sufficiently incompressible that it can transmit the faint pulse of renewed demand as a pencil loses graphite to paper. It does this in a lossless manner, so that the faintest pulse permeates every level of a global structure of production.

With electronic goods such as computer software and data, this hydraulic system is filled with gas, not an incompressible fluid. No, it's worse even than that. It's more like a perfect vacuum that is utterly incapable of communicating any signal at all.

To see this, consider a programmer who builds a reusable software component and allows someone else to build it into a larger application. This is normally done with the understanding that no royalties will be paid in proportion to how many copies of the higher order are sold. Thus regardless of whether the higher-order vendor is successful or unsuccessful, neither information nor revenue passes down the hierarchy. As the top-level vendor replicates copies of his product for sale, the inner component is automatically and invisibly replicated within it. This creates a vapor lock that blocks the lower level from receiving either revenue or information as to the utility his product has provided to his customers.

Regardless of the demand at the consuming end, and regardless of the price consumers might be willing to pay, neither information nor revenue passes to the supply end. The entire structure of production that served the tangible goods providers so completely becomes utterly useless for those who would provide electronic goods.

It is important to see that the breakdown here is not confined to any particular level of this system. In particular, this is not a problem that can be addressed at just the very top of this chain. There is no shortage of these unigranularity solutions. The software industry is awash in solutions for the only level of this hierarchy that is commercially viable already, large computer applications.

Microsoft, for example, has demonstrated complete mastery of one such solution: attaching their electronic property to paper, cellophane, and plastic. This simple expedient allows their goods to be bought and sold exactly like cornflakes and detergent.

But much higher-tech solutions to this same subset of the real problem are emerging everywhere we look these days. These range from secure web servers, to various solutions for unlocking software, however distributed, upon payment of a fee. A few years ago, some vendors even ran short-lived experiments with dongles and other "software protection" devices. This was to prevent losses to customers for whom copyright laws weren't a sufficient disincentive to redistributing software once it has been freed from its industrial-age encumbrances.

The problem with these solutions is that they work only at the top of the structure of the production tree. This is paradoxically the one place in the tree that doesn't have a real problem getting paid, as the success of Microsoft will attest. Addressing the problem at any single level is like trying to address a famine by paying the baker to bake bread faster, while ignoring how to get the farmers and millers to produce more wheat and flour.

Similarly, the Denver International Airport debacle didn't happen because the software vendors had a problem getting paid. They couldn't deliver what they'd promised because nobody was willing to participate in a structure of production for software, comparable to the way wheat farmers and millers work out their entire careers to help the baker produce his bread. Precisely the inverse was true of the engineers who provided the tangible components of this airport. The tangible goods rolled in on time and within budget, be they sophisticated components like the airplanes and buildings to mundane ones like wooden pencils.

It is important to realize the magnitude of this breakdown. This is not merely a dysfunction of the particular set of companies who happened to bid on the Denver Airport software. Nor is it a dysfunction of the larger system we think of as the software engineering community. This was not ignorance, nor disobedience, nor criminality on the part of either customer or provider. The breakdown was on a far more fundamental level than something that can be blamed on individuals, organizations, or industries. It was a breakdown of physics, not society; a breakdown of the physical laws that support the very binding force that make long-range social orders possible. The physical laws that gave mature domains the very qualities that make us refer to them as mature don't work for the enterprises of the electronic frontier.

2.6 PENCILS AS THE FRUIT OF THE PENCIL TREE

In a previous section, I analogized the social order that produces pencils with a vast hydraulic system, filled with an incompressible fluid capable of transmitting minute pulses of demand around the globe. I adopted this machine-like analogy for a living social system because such analogies are widespread today. Indeed, we're so immersed in such rhetoric that we fail to realize that we're speaking of living social systems composed of flesh and blood people, as if they were machines.

This section will introduce an entirely different analogy, one that avoids perpetrating the disastrous illusion that social systems are like machines in any way. To emphasize that the pencil-producing system is a living, evolutionary system, I'll call it the pencil tree. Of course, this tree isn't the kind that grows in an orchard. It is a figurative tree, rooted in human strivings. Its cells are individual people such as the rape seed grower in the West Indies, the lumberjack in the forests of Oregon, and the clerk in the office supplies store in Denver. Its tissues and organs, its roots, leaves, and branches, are businesses, corporations, and firms. And exactly as biological trees produce not just fruit, but also the components needed to produce them, the pencil tree produces not only wooden pencils as its fruit, but a comparably vast array of raw materials and intermediate components. Its formal name is *structure of production.*

Exactly like biological organisms, this pencil tree was never designed. It *evolved* as forest trees evolved in response to their intimate involvement with their environment. The organizing force for this evolution was energy. For biological trees, this energy came from a photosynthetic combination of carbon dioxide, water, and sunlight. For the pencil tree, this energy source is the steady trickle of customers' nickels and dimes. Just as photosynthesis provides the energy that allowed biological trees to evolve, pencil trees evolve from the energy of the commercial exchange transaction.

If you feel uncomfortable with the use of these biological analogies for organizations and economies, you're hardly alone. Most people are far more comfortable using machine-age analogies for organizations and economies. Every day we hear machine-age analogies on the news. We hear that the "regulators" feel that the economy is "overheating" and needs to be "cooled down," or vice versa, or that some monopoly is "out of control" and the "brakes" should be applied. Each of the quoted terms in this paragraph uses terms that highlight the assumption that human systems, are most fundamentally, like machines. And implicitly, that they should be treated like machines.

These machine-age analogies are just as common at the organizational level. The runaway fashion in organizational theory right now is the notion of "reengineering" the organization. This is often coupled with the desire to hire some nerd to "model" the organization, generally involving the implicit assumption that such a model will help central planners (managers) comprehend the complexity of their organization in order to "control" it better, exactly as they might control a machine.

A final symptom of machine-age thinking is our techno-centric habit of thinking that manufactured products, like pencils, are produced by *machines* instead of *people*. The tale of the pencil shows that pencils are no more produced by the milling machine in San Leandro than software is written by computers. The machines are ancillary and peripheral, not central. Production is fundamentally a human process, not a mechanical one. It involves individuals and organizations, using tools as an ancillary, albeit important, part of the human-centric production process.

This criticism of speaking of organizations and economies as machines originates from Michael Rothschild's book, *Bionomics*. He argues, convincingly I believe, that the use of machine-age terminology shapes the way we think about organizations and economies, and that the way we think about them leads to actions that can have devastating consequences in practice.

For example, notice that we think of machines as things that we can stand outside of, to design, understand, and control them. The designer is not immersed in the machine the way we are immersed in and shaped by our organizations. The belief that we can escape from this immersion by pretending that we can engineer organizations the way we do machines is one of the pervasive myths arising from industrial-age thinking that Rothchild is trying to expose.

Rothschild argues that this vocabulary is actively harmful in that it persuades us we understand social systems well enough to succeed. He argues that biological or ecological analogies might calm our ardor for interfering in matters we neither understand nor control. Thinking about companies as organisms and economies as ecologies; as rain forests or oceans and not as machines, might foster the same cautionary instincts that we'd feel if someone proposed to "improve" or "reengineer" the rain forests by spraying DDT to kill off the mosquitoes.

2.7 INDIAN POTTERY

If sophisticated structure of production trees doesn't work for objects-made-of-bits, what explains the incredible growth of the internet? What about the

Figure 2.3 This mural affirms the bonds between man and nature. The bowl represents creativity; corn, the staff of life. —The Fruits of Life by Charles Lovato on display at the Indian Pueblo Cultural Center, Albuquerque.

phenomenal growth of the software industry as a whole? If commerce is so powerful, why should any small-granularity reusable software components exist at all? Commercial incentive could explain my own production of electronic properties such as this book, but what explains the effort I and many others spend producing free material for the web, such as my own Middle of Nowhere web?

Although commerce clearly plays a strong and occasionally dominant role as an organizing force for human society, it is not the only or even the strongest force governing human affairs. This was brought home to me quite recently during a visit to the Santa Fe Institute of New Mexico, when I made numerous side trips to visit several Indian Pueblo communities in that area. Since these questions were fresh on my mind, I took advantage of the opportunity to discuss them with people from a social order entirely different from my own. I spent several hours with several Pueblo Indians, discussing how the "Indian Way" in Fig. 2.3 differs from the "Western Way" typified in the tale of the wooden pencil.

The pot in Fig. 2.4 typifies the black-on-black style of pottery produced at the the San Ildefonso Pueblo near Santa Fe, New Mexico. The pot shown here was produced by Barbara Gonzales, the fifth-generation granddaughter of Maria Martinez. Maria became world-famous for discovering how to make the highly polished, jet-black pottery that has made San Ildefonso's Black Pottery a staple in museums around the world.

Barbara's husband runs the family's pottery shop within the pueblo's walls. This is my best recollection of an extended conversation in which I tried to absorb the differences between the Indian and Western ways:

In the Indian way, spirituality isn't something that we do only on Sunday. It is an integral part of everything we do. When we dig clay from the pit on that mesa over there, we begin by thanking Mother

Earth for providing the clay. We take the clay back to the pueblo, wash it with water from that river over there, and sift it until it looks like this (referring to a sample he keeps on hand for visitors). There are also places where we dig volcanic ash which we add to the clay as a binder.

Once the clay and binder have been sifted, combined in the right proportions with water, and worked until it is pliable, we coil the clay by hand to produce the rough shape of the pot we want. When this dries sufficiently, we scrape away the excess to produce the finished shape.

Then comes the hard part. We grease the pot with lard and polish it with a stone until the pot has a glossy sheen all over. These stones are very important to us. They are usually inherited from mother to daughter. Then we scrape away the gloss into a decorative Indian design such as these eagle feather or water serpent examples.

The very last step is to fire the pot, after which no further processing or decoration can be done. We go out to the pastures and gather dried cow and horse manure. We set the unfired pots right on the ground over there and cover them with tin (license plates) to protect them from the ashes. If we're building reddish pottery (referring to a glossy reddish pot on a shelf), we simply pile dry cow cakes over the license plates and set it alight. But if we're building the black pottery that Maria discovered, we add horse manure to the fire. Maria's big discovery was that horse manure makes the difference between red and black pottery.

— Robert Gonzales, San Ildefonso Pueblo, N. Mex.

But what, you may well ask, does Indian pottery have in common with computer software? The answer is that they have absolutely nothing in common at all. Not if it's pots and software that we've laid on the stage of our microscope. At every level of magnification, pots are entirely different from computer software. At the gross level, pots are visible and software is invisible. They are produced from entirely different subcomponents, using technologies that are even more different. It is hard to imagine anything more low-tech than the Indian approach to pottery building, in which every step from digging the clay, to coiling and shaping and polishing the pot, to firing it in an open fire, using cow and horse dung, involves technologies that date back to antiquity. Conversely, computer software is one of the highest-tech activities imaginable.

Figure 2.4 This beautiful pot was created by Barbara Gonzales of the San Ildefonso Pueblo. Much of the pottery produced here is entirely black, with a black-on-black design incised into a highly polished surface. Firing iron-rich clay in an oxidizing or reducing atmosphere produces pots that are entirely red or jet black. The pot shown here has both of these colors due to careful control of the oxidation/reduction process. — Pot designed and created by Barbara Tahn-moo-whe Gonzales. Photograph by Karen Strom; used by permission.

At the highest magnification imaginable, they are different in the most fundamental way possible. Pots are made of atoms whereas software is composed of bits. The former are hard to replicate and transport while software can be replicated and transported at the speed of light.

This way of comparing them originates from a techno-centric view that emphasizes technical processes over social processes. Make the same comparison in human-centric terms and the similarities become painfully obvious. The social order that these Indians used to produce pots is exactly the same as the social process that the Denver Airport software developers used to build their software.

Figure 2.5 depicts the "Indian Way" for producing pottery. At the bottom is Nature, the origin of clay, water, and fire used in making these pots. At the top are the end users, tourists such as myself. The circles represent the economic actors in this system, either individuals or family-based teams (Mr. Gonzales wasn't specific as to what extent family members cooperate in making and selling their pottery). Every participant fabricates everything needed from first principles. This figure should be contrasted with the spe-

Customers

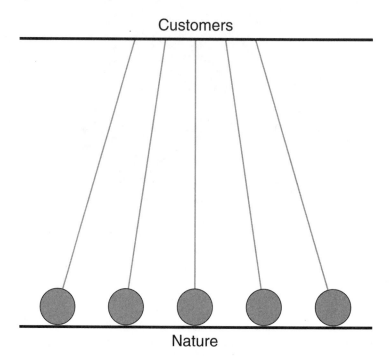

Nature

Figure 2.5 This figure eliminates all details of Mr. Gonzales' description of how pottery is produced by the Indians of San Ildefonso Pueblo to highlight the basic skeleton of the social order. At the bottom of the figure is nature, the source of clay, water, and fuel. At the top are end users.

cialized structure of production of Fig. 2.2 that would allow products to be assembled from markets in prefabricated components.

Although no visitor could overlook the degree to which these Indian communities have been influenced by commercial interactions with the Santa Fe tourist industry, the Indians try to confine these interactions to the periphery of their society and resist adopting it internally. When I asked Mr. Gonzales how his community would react if he offered to sell clay to his neighbors to save them the trouble of climbing the mesa to dig it themselves, he made it perfectly clear that this would be totally unacceptable. Although the commercial interchange that dominates our social order is not unheard of in such communities, it is accepted only reluctantly and involves an affront to very deeply held beliefs.

This example helps to understand the Denver International Airport debacle, and by extension, the software crisis as a whole. The social order of the San Ildefonso pueblo is capable of supplying pottery for the tourists while satisfying their own internal creative and cultural needs. But if they had been asked to supply luggage conveyor systems, airplanes, and runways for the Denver airport, or even mass-produced vessels for a Coca-Cola bottling plant, they'd have encountered the same breakdowns that the software vendors did. The reason has nothing to do with technology, for they'd have encountered these breakdowns *even if suitable manufacturing technologies were externally provided.* These Indians organize their lives into a social order exactly like the one the software engineering community uses. And although artistic social orders are capable of astounding achievements in creativity, they're unable to handle the overwhelming complexity of mass production, complexity that industrial social orders handle so efficiently that they make complexity seem routine.

2.8 ELECTRONIC PENCILS

Modern computer/communication systems are among the most complex technologies ever assembled by human beings... to be able to understand, predict, and control the behavior of these technologies requires a powerful theory of complex processes. No such theory yet exists, although it remains a major goal of computer science.
— 1985 Office of Technology Assessment Report

I have been a committed Microsoft Word fan since I switched from DOS to Mac machines in the late 1980s. I was really excited by the features of Word 6: integrated grammar checker, revision control, PageMaker-like kerning, real DOS compatibility.

Then I started using it about six weeks ago. My PowerBook runs at 33 MHz, yet this monster word processor crawls along more slowly than any I've used in fifteen years. It's not as if I'm doing complex rendering. The neat-sounding features turn out to be bugs. It thinks for me in ways I wish it didn't. And, fully configured, it takes almost 10% of my hard disk. And, it loses files. And, transferring files to a DOS Word 6 machine doesn't work. And, and, and...

I wonder if Microsoft understands the consequences of doing this to me? Besides crimping my personal productivity, I've become an unhappy customer. Once dissatisfied, I started noticing the bugs that I'd never have seen. I've already told my average fourteen friends about how crummy it is. Plus, I'll be much slower to buy the next product. I'm unhappy.

> *Ordinarily, I'm so delighted with Mac software that I act as a regional missionary. I have the zeal of a religious convert. In the last eight years, I've standardized four organizations on Macs with Microsoft products. I'll have to think about it the next time.*
> — essay by jrsumser@well.sf.ca.us at
> http://magnet.mednet.net/zisk/DaveNet/DW59.html, used by permission.

Now let's look at a third and final example from the field of computer software. Since I don't have access to the vendors who built the Denver International Airport software and since baggage cart software won't be readily understood, we'll use word processor software instead. The particular example is Microsoft Word 5.1a for the Macintosh. I chose this example because Microsoft products will be familiar to most readers, and because Word is one of the tools I used to write this book.[3]

The first symptom of a major difference between wooden pencils and these electronic pencils is the number of people involved in producing them. Whereas thousands of people work to bring wooden pencils to market, according to the manager of this project, this version was developed by a team of only eight programmers.[4] This number includes only the programmers who worked hands-on in developing it, and excludes those who played an indirect role in testing, documenting, selling, and marketing it. The point is that the full internal complexity was such that it could be managed by the eight flesh-and-blood human beings who had hands-on access to the source code.

Of course, we could easily enlarge the frame to view these eight programmers within the advanced industrial structure of production that we call "the computer industry." The programmers of Microsoft clearly didn't mine their own silicon to fabricate their own computers. They relied on Silicon Valley to supply them with these and focused on building the software. However, this obscures the distinction I'm trying to illuminate. The computer industry builds tangible goods fully comparable to the wooden pencil. Computers and pencils can be bought and sold as tangible goods made of atoms. Thus the prerequisite for evolving a robust social order is fulfilled, a robust energy source around which industrial structures of production can evolve.

---

[3] I actually have several text editors of various degrees of sophistication that I use for different tasks. Although Word is the best known, I've used it very rarely, using Nisus for most work because of its comprehensive string search capabilities.

[4] I had called him to see what impact object technology was having on the ability to coordinate larger software development teams. He said he thought that object-oriented programming languages, C++ in this case, might allow him to double team sizes in future versions. Although this is clearly an improvement, object technology alone is not going to get us to the level of cooperation revealed in the tale of the wooden pencil.

Here lies the paradox that will concern us in this book. We believe very deeply, and quite correctly in some sense, that software products are far more complex than something so mundane as a wooden pencil. But here we've seen that the Microsoft electronic pencil involved a mere handful of programmers, but thousands of people needed to cooperate to make the complexity of a wooden pencil manageable.

2.9 SOFTWARE ICS

In 1982 I cofounded a software company, Stepstone, whose goal was to play the role for software that Intel does for computers. Its main product was a large-granularity software component, a compiler for an object-oriented programming language called Objective-C. Its other products were smaller-granularity components that we called *Software-ICs*. We chose this name as an explicit analogy with the silicon integrated circuits that made Intel one of the giants of American industry.

The Objective-C compiler is a tool for binding Software-ICs to each other, just as a soldering iron is used to solder chips to circuit boards. The compiler let our customers assemble our small-granularity components into larger applications instead of having to fabricate everything from first principles. Software-ICs are objects made of bits. They are generally much smaller than computer applications, too small to be useful by themselves. They are used in building larger-granularity applications in precisely the sense that silicon chips are used to build computers and factice is used to build pencils.

Although this company is still in operation today, it never achieved anything like the prosperity or influence of Intel. Figure 2.6 shows why the hardware industry has managed to achieve an industrial structure of production by building upon the shoulders of companies like Intel, whereas the still-dominant practice in software is analogous to the fabricate-from-scratch practices of colonial era manufacturing before the industrial revolution.

In industries that produce tangible goods the relationship between Intel and its customer is shown by the slanted line in this figure. When the relationship begins, the customer is in the development stage. Its sales are small at this point, but so are its expenses. Its payments to Intel are small, just enough to buy a few components to build breadboard circuits for testing the design. If the product fails at this point, little is lost. But if the customer prospers, so does Intel, and in direct proportion to its customers' success.

Contrast this relationship with that of Stepstone and its customers. The Stepstone product was a collection of small components bundled as a collection, or ICPak, that was large enough to justify advertising, marketing, and distributing it as a commercial product. The price of this collection is

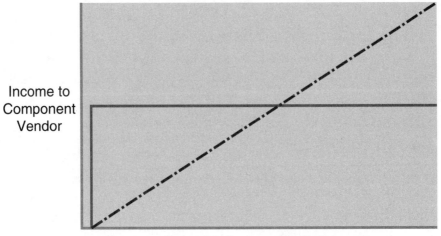

Income to Component Vendor

Sales by Component Consumer

Figure 2.6 This figure relates income to a component vendor, which varies according to the sales of its customers. The sloping line applies to tangible components such as carpentry nails, farm products, and silicon chips. Since these are sold by the copy, a component user's expenses are directly proportional to sales. The stepped line applies to goods that can be replicated locally by the customer. The initial cost is higher, which limits acceptability to small struggling customers. Yet the overall income may be less, which limits the vendor's incentive for providing such goods in the first place.

shown by the red line, which reflects the underlying assumption that "owning" software means acquiring the bits along with the right to replicate as many copies as one needs.

Now contrast what this pricing model means to two cases: a struggling startup and a large successful company. Since the large company can expect to replicate copies of the Stepstone product within large numbers of their own, this fixed price would seem like a very good deal. But on the other hand, a small and struggling but creative garage shop operation might have a hard time justifying this price. We were continually faced with a quandary that the rape seed oil farmer doesn't have, of being unable to strike a fair balance between large and small consumers of our goods. In fact, there is no way of striking this balance so long as the only viable basis for commerce is to sell master copies from which customers can replicate however many copies they might need.

With tangible goods, a component vendor's revenue is strictly proportional to the customer's ability to sell goods based on those components. The miller has every reason to hope that the baker will succeed at selling his bread, since bread sales directly affect the miller's sales of flour. Intel has every reason to hope that its customers will prosper because their prosperity is strictly determined by the number of computers its customers can sell.

But with electronic goods, since the vendor's revenue is confined to the first sale, the vendor has no such interest in his customer's success. And since revenue must come entirely from the first sales, the vendors are forced to set this initial price so high that companies who are already successful can afford to buy them.

Second, new customers who are uncertain of their prospects are unable to afford the higher price of entry. Small struggling companies can easily afford one or two Intel chips for breadboarding, knowing full well that they'll need to pay the same amount for every product they ship. It seems fair to them that they'll pay far more in the long run, because they incur this cost only if they're successful. But a small software operation might expect to sell a few dozen copies of a product while Microsoft might sell millions. If they're both charged the same amount to replicate a component of these products, this overcharges the struggling small company and undercharges the large successful one.

Markets in reusable software components have never evolved, not because of techno-centric reasons, such as shortcomings of tools, languages, or methodologies. They fail for human-centric reasons; i.e., economics. The ease of software replication requires component vendors to overcharge those who can least afford to pay while simultaneously undercharging the very group to whom their components have made the greatest contribution.

2.10 SUMMARY

This chapter has described a breakdown in the commercial exchange transaction mechanism that has bound individuals into mutually supporting social enterprises since antiquity. The commercial exchange transaction, which is itself based on physical laws such as conservation of mass, has effectively mobilized individuals and organizations to cooperate with the vast structures of production in spite of the chasms of language and culture that would otherwise separate them. The commercial exchange transaction plays the same role for social systems as the long-range binding forces that bind planets and suns in their orbits in the skies.

This mechanism breaks down for the easily copied goods of the electronic frontier. Conservation of mass doesn't apply to goods made of bits. Thus individuals and organizations lack the incentive to cooperate within a structure of production for larger granularity electronic products.

This diagnosis is actually an optimistic conclusion since it directs attention toward the source of the problem, a breakdown in incentive mechanism that has prevented comparable structures of production from evolving. Since conservation of mass doesn't support the pay-to-acquire basis of ownership of tangible goods, we must provide something else that could support incremental revenue flows to those shown in Fig. 2.6. Chapter 6 will describe such an infrastructure in which revenue flows throughout a structure of production are based on *uses* of objects and not on acquisition of *copies*.

Chapter 3

Software Crisis

Of all the monsters that fill the nightmares of our folklore, none terrify more than werewolves, because they transform unexpectedly from the familiar into horrors. For these, one seeks bullets of silver that can magically lay them to rest. The familiar software project, at least as seen by the nontechnical manager, has something of this character; it is usually innocent and straightforward, but is capable of becoming a monster of missed schedules, blown budgets, and flawed products. So we hear desperate cries for a silver bullet... something to make software costs drop as rapidly as computer hardware costs do.

—Fred Brooks, *No Silver Bullet; Essence and Accidents in Software Engineering,*
Computer Magazine; April 1987; and Information Processing 1986,
ISBN No. 0444-7077-3, Elsevier Science Publishers B.V. (North-Holland)

The term, software crisis, was coined during the 1967 NATO Software Engineering Conference, nearly a quarter-century ago. During the very period that manufacturing has achieved notoriety for delivering exponentially growing price-performance, software remains intractable, immune to the organizational innovations that work so well in other domains.

In one of the most influential software books of this era, *The Mythical Man-Month,*[1] Fred Brooks observed that adding more people to a late software project only makes matters worse. And in a subsequent article, *No Silver Bullet; Essence and Accidents of Software Engineering,* he argues that the difficulties are inevitable, arising from the inescapable essence of software and not from 'accident', something we're doing wrong in its production.

3.1 THE GOOD NEWS: THERE IS A SILVER BULLET

Lord, please let me find a one-armed economist so we won't always hear "On the other hand..."

— Edgar R. Fiedler

---

[1] Frederick P. Brooks, Jr., *The Mythical Man-Month*, Addison-Wesley, 1978. ISBN 0-201-00650-2

Joseph Joubert once said, "The aim of argument should not be victory but progress." Although I'll open this chapter by appearing to disagree with Brooks' dire conclusion, it will soon become clear this chapter is a serious attempt to absorb Brooks' insights, build upon them, and advance even further. But in the manner of Fiedler's one-armed economist, I'll start by painting what will ultimately prove to be a complex picture as the oversimplified black versus white outline.

The phrase, *No silver bullet,* turns up so often in technical discussions that it clearly holds some deep meaning for us. But precisely what meaning? Does it mean nothing more than that the solution won't be as easy as adopting the latest whiz-bang to come down the pike? If so, the phrase is nothing more than a truism. Tools alone are clearly useless without effort from those who use them.

Furthermore, silver bullets certainly do exist. The Lone Ranger used to hand them out to bewildered strangers before riding into the sunset with his trademark, "Hi ho, Silver!" And they are just as common today. Winchester Silvertip™ Hollow Points are the best-selling ammunition of today's urban killing fields. A quick scan with the Yahoo search tool even shows that an enterprising arms merchant is setting up shop on the web under the name Silver Bullet Law Enforcement Products.

Brooks is actually claiming far more than either of these interpretations. The subtitle of this article, *Essence and Accidents of Software Engineering,* means to convey that the software crisis *cannot* be solved, *not even in principle.* If the software crisis really originates from the inescapable essence of the software and not from something we're doing wrong in building it, what hope could there possibly be?

But, as we look to the horizon of a decade hence, we see no silver bullet. There is no single development, in either technology or in management technique, that by itself promises even one order of magnitude improvement in productivity, in reliability, in simplicity. In this article I shall try to show why, by examining both the nature of the software problem and the properties of the bullets proposed.

Skepticism is not pessimism, however. Although we see no startling breakthrough—and indeed, I believe such to be inconsistent with the nature of software—many encouraging innovations are under way. A disciplined, consistent effort to develop, propagate, and exploit these innovations should indeed yield an order of magnitude improvement. There is no royal road, but there is a road.

> *The first step toward the management of disease was replacement of demon theories and humours theories by the germ theory. That very step, the beginning of hope, in itself dashed all hopes of magical solutions. It told workers that progress would be made stepwise, at great effort, and that a persistent, unremitting care would have to be paid to a discipline of cleanliness. So it is with software engineering.*
> — Fred Brooks

In other words, Brooks found no solution in "the bullets proposed" and "a decade hence." To say the same thing in the terminology I shall develop in this chapter, Brooks found no solution *within the present-day paradigm,* the way we develop software today.

This is the origin of the apparent disagreement. An optimist and a pessimist are debating whether the glass is half-empty or half-full. Brooks is saying there's no solution, even in principle, *so long as the present paradigm of software development is unchanged.* I'll argue that there *is* a solution, but that it involves changing the paradigm that has governed the electronic frontier since its inception.

Prison administrators recognize two different groups in convict populations. One views the prison walls as immovable obstacles. They either serve out their terms in hopeless despair or they collaborate, cooperating with the oppressive system to co-opt it to their personal advantage. The other group views the very same walls, the same bundle of facts, as an irresistible force, the motivating factor for an urge to escape that even the strongest walls cannot withstand for long. The same is true of tall mountains. To most of us, they are immovable obstacles. But mountain climbers see them as irresistible forces, obstacles to be surmounted simply because they are there.

3.2 THE BAD NEWS: SILVER BULLETS ARE PARADIGM SHIFTS, NOT TOOLS

The paradigm shift that I'll be advocating in this book is not a new idea. It involves adopting the economic engine that fueled the phenomenal progress in mature engineering domains such as computer manufacturing. The secret of their success was to divide up the work among different companies, each of whom supplies well-understood parts for others to assemble, with incentives provided through commercial exchange transactions. Our challenge, which will be addressed in later chapters, is that commercial exchange transactions don't work well for goods that can be replicated and transported at the speed of light.

3.3 THE COPERNICAN REVOLUTION

*Logic and reason sprang from the use of angles and lines. These tools
became the basic instruments of Western thought: indeed, Aristotle's
system of logic was referred to as the Organon (the tool). With it we
were set on the rationalist road to the view that knowledge gained
through the use of the model was the only knowledge worth having.
Science began its fight to supplant myth and magic on the grounds
that it provided more valid explanations of nature.*
— James Burke, *The Day the Universe Changed*

The magical solutions to the seemingly insurmountable obstacles of the
past originated from scientific revolutions, silver bullets for major breakdowns of the past. They were not primarily about technology, i.e., better
tools for doing things the old way, even though tool improvements generally
followed in their wake. These silver bullets were paradigm shifts; not tools,
but new ways of looking at the problem.

This philosopher-centric paradigm reflected in this quotation dominated
mankind's view of its place in the universe from Plato until the end of the
Middle Ages. It was responsible for the medieval emphasis on logical techniques such as Aquinas' advocacy of syllogism for arriving at a deeper understanding of the true nature of God. To medieval alchemists with more
earthly interests, syllogism was the rational underpinnings behind their
search for the philosophers' stone, a way of transforming base metals such
as lead into gold.

A famous outcome of this emphasis on logic over experimental observation was the Aristotelian and Ptolemaic cosmologies of the universe. This
cosmology is usually traced back to Aristotle, but is based on what peasants
could see for themselves. The sky seems to move. The stars wheel past every
night. At the poles they never disappear, circling round the North Star. Five
wandering lights, or planets from the Greek word for 'wanderer', move
among the fixed stars. The moon circles the earth, as does the sun.

Incredibly, even though Aristarchus, the head of Aristotle's Lyceum, had
published his belief in a heliocentric system and even tried to estimate the
distances and sizes of the moon and the stars, his theory was never accepted
by the Greeks. The Greeks preferred Aristotle's explanation that these and
other phenomena existed by means of a system of eight crystalline spheres
on which the sun, moon, planets, and stars were fixed. Each sphere revolved
around the earth, which never moved. In the middle, at the center of God's
vision and of course the primary subject of his concern, was mankind, the
church, and the ancient philosophers. In the same way, software developers

are at the center of the software universe, with users and their representatives, the managers, revolving around outside, in the margins of this frame.

> *Plato examined the difference between the untrustworthy and*
> *changing world of the senses and that of the permanent truths which*
> *were only to be found through rational thought. The unchanging*
> *elements of geometry were the measures of this ideal, permanent*
> *thought-world with which the transitory world of everyday existence*
> *could be identified, and against which it might be accessed. The union*
> *of logic with geometry laid the foundations of the Western way of life.*
> — James Burke, *The Day the Universe Changed*

The collapse of this cosmology was triggered by something casual observers could look up and see for themselves. At times, the planets seemed to change course. Mars, in particular, seemed to stop and even to recede at times, the so-called retrograde motion of Mars. Given Aristotle's system of celestial spheres that could not easily change their direction, this was a serious problem indeed.

Claudius Ptolemy saved the day by developing an acceptable patch. He proposed that the difficulty could be resolved if the planets were attached to smaller secondary spheres, called 'epicycles', each affixed to the main sphere. Although this was clearly a patch, it was highly 'successful'. Astronomers found that they could adjust the patched model to account for almost any observation by simply postulating as many additional epicycles as they might need. But eventually it became clear that the complexity of astronomy was increasing more rapidly than its accuracy and that a discrepancy corrected in one place was likely to show up in another. By the beginning of the sixteenth century, astronomers needed *ninety* of Ptolemy's epicycles to compute reasonably accurate calendars; by now a true sense of crisis had developed.

Just as with the software crisis, the complexity of patching a paradigm instead of replacing it had festered, bred, and proliferated. Complexity was entirely out of hand, exactly as in software today. It is easy to imagine one of these early astronomers, driven to wits' end by unceasing complexity, venting his frustration and despair by writing an article exactly like Fred Brooks'. It might well have been titled *No Silver Bullet; Essence versus Accidents in Astrophysics.*

No doubt this author's peers would have reacted as programmers react to Brooks' article today. Clearly no mere technology could possibly eliminate this mind-numbing complexity. Or more to the point, no mere technology

could possibly eliminate society's need to maintain a powerful establishment of astronomers in order to carry out this essential, but intrinsically
complex, task.

So in 1514 the secretary to the Pope asked an unknown Polish priest and
mathematician, Copernicus, to look into calendar reform. And during the
following centuries, Copernicus, Galileo, Kepler, Descartes, and many others
forged what we today can see was nothing less than a silver bullet for the
astronomy crisis.

Exactly as Brooks argued in his article, this silver bullet was not a new
technology. Copernicus didn't invent a new calculus, slide rule, or computer
that his peers might use to compute epicycles faster. His innovation was far
more encompassing than any mere technology. The silver bullet was Copernicus' paradigm shift of moving the axis of heavenly motion from the earth
to the sun.

In hindsight it is obvious that Copernicus' heliocentric cosmology was
anything but magical. But as Arthur C. Clarke said, "Any sufficiently advanced technology is indistinguishable from magic." A modern astronomer's ability to compute in moments what would have taken the ancient
astronomers months would have seemed magical indeed.

The good news was that Copernicus had discovered a silver bullet for the
astronomy crisis. The bad news was that this silver bullet was a paradigm
shift, not a technology. This proved to be very bad news indeed for the entrenched establishment of his day. The scientific revolution that he triggered
eventually displaced the theologians (Martin Luther called him an "upstart
astrologer"), philosophers, and alchemists of his era with the engineers, scientists, and technologists who have retained their dominance until today.

The point of all this is to show that paradigm shifts, as do all revolutions,
have extremely disagreeable implications to incumbent establishments:

- They clash with deeply ingrained beliefs and value systems.

- They are chaotic, disruptive, and are invariably regarded as evil.

- They take a long time (a generation or more).

- They never benefit bottom line immediate profits.

- They overthrow entrenched establishments and bring disenfranchised
  groups to power.

This warning is to compensate for the way this book is organized. The
first half is devoted to the good news. It explains what the term, software industrial revolution, means and in what sense it can be a silver bullet for the

software crisis. The second half will drop the other shoe by considering the magnitude of the changes that will be involved in making the software industrial revolution a reality.

3.4 SOFTWARE COMPLEXITY

Overseeing programmers is a managerial challenge roughly comparable to herding cats.
— Washington Post Magazine, June 9, 1985

A physician, a civil engineer, and a computer scientist were arguing about what was the oldest profession. The physician remarked, "In the Bible, it says that God created Eve from a rib taken out of Adam. This clearly required surgery, and so I can claim that mine is the oldest profession." The civil engineer interrupted and said, "But even earlier in the book of Genesis, it states that God created the order of the heavens and the earth from out of the chaos. This was the first and certainly the most spectacular application of civil engineering." The computer scientist leaned back in her chair and said, "And, but who do you think created the chaos?"
— Grady Booch, *Object-oriented Design with Applications*;
Benjamin Cummings Publishing Company; 1991

Brooks' article provided a list of qualities that make software essentially different from hardware. I'll discuss his entire list later after adding a few that he neglected to mention. But for now, let's concentrate on only the first item in Brooks' list, complexity, as a way of showing the change in viewpoint that terms like "paradigm shift" and "software industrial revolution" mean to convey.

Is software inherently, i.e., essentially, complex? Or is complexity an accident, a consequence of the way we build software today? The software engineering literature overwhelmingly adopts the former assessment. For example, in *Object-oriented Design with Applications*, Grady Booch devotes the first chapter to complexity, endorsing Brooks' assessment almost verbatim. For example, the very first paragraph:

A dying star on the verge of collapse, a child learning how to read, white blood cells rushing to attack a virus: these are but a few of the objects in the physical world that involve truly awesome complexity. Software may also involve elements of great complexity; however, the complexity we find here is of a fundamentally different kind. As Brooks points out, "Einstein argued that there must be simplified explanations of nature, because God is not capricious or arbitrary. No such faith comforts the software engineer. Much of the complexity that he must master is arbitrary complexity."

— Grady Booch

The second section of the same book begins with the following dissection of the origins of the complexity of software:

As Brooks suggests, "The complexity of software is an essential property, not an accidental one." We observe that this inherent complexity derives from four elements: the complexity of the problem domain, the difficulty of managing the development process, the flexibility possible through software, and the problems of characterizing the behavior of discrete systems.

— Grady Booch

These quotations demonstrate what I mean by the existing paradigm. Software certainly *is* complex—the way we build it today. Anything would be complex that is built by having every individual design and fabricate everything from first principles.

Programmers overwhelmingly agree that software is more complicated than home plumbing. But is the difference really *essential* in the sense that Brooks and Plato used this term? Isn't the complexity of software actually an artifact of the way we build it? Imagine the complexity of an ordinary plumbing system if plumbers built plumbing systems the way we build software. Each plumber would need to be a geologist to know where to dig mines. He'd need to be expert in refining and milling in order to refine ore into raw materials. He'd need to invent, design, build, and test the vast multitude of parts that make up even the most mundane of plumbing systems. And finally he'd need to be a plumber to do what his customer hired him for in the first place. When all this is finished, what is his customer's perception of it likely to be? Probably that it was too expensive, it took too long, and it was still defective, just what we expect of software today.

Since software is hand-crafted, fabricated from first principles and not assembled from prefabricated components, it does not benefit from the organizational innovations that made plumbing an unremarkable activity today.

Suppose that this plumber adhered to the most advanced process-centric dogmas that we look to for the salvation of software today? Suppose that our plumber were a certified Level 5 plumber according to the process maturity model that DARPA and SEI are using to certify developers for the Department of Defense?[2] Wouldn't even the most aggressive application of process maturity certification, object-oriented technologies, CASE tools, formal methods, inspection teams, configuration control systems, and analysis/design methodologies seem irrelevant to the common sense concern that something far more basic is wrong with our plumber's whole approach?

3.5 SOFTWARE INDUSTRIAL REVOLUTION

Today's plumbing market arose from innovations pioneered during the industrial revolution. One of the innovations of that revolution was the concept of standard parts, prefabricated, interchangeable components that we accept today as routine. Specialists in geology, refining, milling, manufacturing, and so forth, were enabled by specialization of labor to spend their entire careers producing ready-to-use components for the plumbing supply market. The plumber is the specialist at the top of that vast chain, enabled to spend his entire career at a single level of this hierarchy, assembling standard, interchangeable plumbing components to produce customized, one-of-a-kind solutions for his customers.

The software industrial revolution is a paradigm shift, a change in belief as to which exemplar is 'best' for thinking about a problem. The dominant paradigm of today is that the best exemplars for thinking about software are abstract (mental) activities such as writing, logic, or mathematics. Since abstract activities are notoriously hard to do by committee, these activities are dominated by fabrication of abstract products from first principles. They rarely proceed by assembling prefabricated parts.

The paradigm shift to be explored in this book involves treating software, not as an abstract activity of individuals, but as a concrete activity of organizations. From this viewpoint, the complexity of software is hardly an essential outgrowth of its unchangeable essence. It is instead an accident, an outgrowth of the fabricate-from-scratch approach encouraged by the established paradigm.

If this paradigm could somehow be changed, organizational principles such as division of labor could allow most of us to work at higher levels of a specialized labor hierarchy by assembling the software we need from prefabricated components. The fabricate-from-scratch construction needed to

---

[2] Watts S. Humphrey, *Managing the Software Process*, Addison-Wesley Publishing Co., Reading, Mass., 1989. Peter H. Feiler and Watts S. Humphrey, *Software Process Development and Enactment: Concepts and Definitions*, Software Engineering Institute, Carnegie Mellon University, Pittsburgh, Pa., 1991. Watts S. Humphrey, *Session Summary: Review of the State-of-the-Art*, Proceedings of the Fifth International Software Process Workshop, Kennebunkport, Maine, 10-13 October 1989, IEEE Computer Society Press, Los Alamitos, Calif., 1990.

build the lowest-level components could then be provided by a specialized and talented minority who staff the foundation levels of a specialized labor hierarchy.

This involves a completely different cosmology for the software universe. The universe revolves, not around the programmer, but around a *market*, a place where producers and consumers exchange prefabricated software components.

This is no longer the programmer's familiar discrete world in which everything is on or off, black or white, either absolutely right or totally wrong. Markets involve continuum questions, not discrete ones. Everything involves fine distinctions between shades of gray, and issues of *tolerance* abound. Is this product tolerably close to what I want to buy? Is it tolerably fast? Tolerably cheap? Tolerably small? Tolerably reliable? Is it tolerably compliant to some vendor-independent specification that I can use it with products from other suppliers?[3]

3.6 THE INTANGIBILITY IMPERATIVE

Testing can only show the presence of defects, not their absence.
— Edsgar Dijkstra

The scientific revolution is today triumphant in all fields but ours. The belief that the internal world of the mind (logic and mathematics) should dominate the external world of the senses (experimentation and testing) continues to dominate academic thinking in this last refuge of the Ptolemaic ideal, those systems of dogmatic belief we call computer 'science' and software 'engineering'.

The established paradigm within these fields is hard to state concisely. This is not because it is subtle and hard to see, but because it is so pervasive that we take it for granted. It will be easier to begin by pointing to specific examples of the established paradigm in order to demonstrate why one way of characterizing the established paradigm is the intangibility imperative: the view that software is primarily about abstract data types, the mental creations of solitary individuals. The opposing view is the tangibility imperative: the view that software should be thought of as concrete data

---

[3] Whether software is best thought of as discrete or continuous will arise again in the narrower context of object technologies in general, and of object-oriented technologies in particular. I only mention this in footnote and won't develop this thought here in order to avoid implying that the software industrial revolution is somehow connected with object-based and/or object-oriented programming technologies. If software industrial revolution is the destination, such technologies are the vehicles. They work just as well for moving ahead or for going around in circles even faster.

types, the encapsulated product of collaborative human labor, boxed and shrinkwrapped to be bought and sold in some information-age equivalent of a store.

The paradigm reflected in the quotations at the beginning of this section is similar to the established paradigm in Copernicus' day. For ancient Greek philosopher and modern computer scientist alike, the following obtain:

- Both center on the intangible abstractions of the world of the mind, around logic and mathematics, with experimental observation of nature somewhere out in the distance. And just like Plato and Aristotle, computer scientists aspire to bringing ever more logic and mathematics to bear on the abstract activities of software design and implementation. The nature-centered orientation of the post-Copernican scientists, revolving around experimental observation of pre-existing components, is not the center of focus. Our tool-building energies concentrate on tools for designing and implementing things from first principles. Specification tools—tools for capturing the specification and then compiling them into tests that determine whether the specification has been met—receive little attention compared to the energy that is devoted to programming languages.

- Both paradigms concentrated power with an elite priesthood, those with the mental prowess to work effectively with intangible abstractions. The majority are left in the role of the astronomers' pragmatic peasant, unable to understand or contribute to the process of calendar construction, only capable of seeing with their own eyes that something is wrong. The customer and his representatives, management, are enfeebled by this abstract process, unable to contribute to the software development process.

The established paradigm is eloquently stated in the following quotation:

> *"I hold the opinion that the construction of computer programs is a*
> *mathematical activity like the solution of differential equations, that*
> *programs can be derived from their specifications through*
> *mathematical insight, calculation, and proof, using algebraic laws as*
> *simple and elegant as those of elementary arithmetic."*
> — H. Fetzer, quoting C.A.R. Hoare in "Program Verification: The Very Idea,"
> Comm. ACM, September 1988

This argues quite explicitly that the appropriate viewpoint is the Aristotelian emphasis on abstraction, not the concrete viewpoint of the post-Aristotelian scientists. The goal is to make software even more of an abstract enterprise, for example by adopting notations based on mathematics and logic as distinct from the procedural programming languages of today. This unapologetically proposes that those with highly developed abstract reasoning skills should stand in the center of this abstract universe, engaged in solitary, intangible, intellectual, creative activity like the theologians, logicians, mathematicians, and philosophers of Aristotle's day.

Although Hoare advocates formal languages like Z or VDM (Vienna Definition Method) in preference to conventional languages like Ada or C, his focus is essentially conventional, focused on the software process, the languages and/or methodologies to be used in fabricating software from first principles. There is no sign of the tangibility imperative of the post-Aristotelian scientist and that of the modern-day engineer, focusing on the externally observable characteristics of the product itself, regardless of the languages, methodologies, and other software development processes that were used to produce it.

Computer scientists are not the only adherents to the established paradigm. The following quotation, by one of Apple's most creative programmers, uses quite a different exemplar, novel writing, to argue that programming should be viewed as a solitary, mental, abstract activity of an individual skilled craftsman, as opposed to an organizational activity like building an airplane:

> *"Reusing other people's code would prove that I don't care about my work. I would no more reuse code than Ernest Hemingway would have reused other authors' paragraphs."*
>
> — Anonymous

This programmer clearly occupies center stage, right in the center of his universe, resisting efforts by other programmers, but especially customers or managers, to influence the outcome. This programmer is saying, "Perhaps there are obscure corners of the universe in which the software crisis is a problem. To me the software crisis means job security." The value system is that of the cottage industry craftsman. The role of customers, and especially of managers, is to stand out of the way, checkbook in hand, admiring the brilliance of this programmer's skill and devotion to his craft.

I copied the following quotation from a heated netnews discussion in which a group of programmers were discussing the ideals of their craft:

> *Engineering is a creative act. By that I mean that you have the power to determine the quality of the products that you design. You also have the power to decide what should be designed and what should not be designed. Both are creative undertakings, each in its way.*

To be an engineer is not to abandon the attributes that distinguish humans from the machines that humans use. To grind out code for the purpose of saying "it is done" is to be a pulp novelist. It is to be an architect of the "functional" buildings that ultimately subvert even their ostensible functions.

For every line of code that you write, if you cannot be proud of it, then you are cheapened; if you can be proud of it, then you are elevated.

Do not be led to forget this, and do not let management tell you what to be proud of. Expect to maintain pride in what you do and satisfy management's needs at the same time. If you cannot do this, then it is management who has imposed the wrong position on you, for the engineer with unrealistically high standards is mythical. If you discover that engineering is not yours to do, find another field.

And when you become a manager, do not let expedience cause you to destroy those qualities that make your best engineers valuable to you and good at what they do. If your wish is to hire a machine, purchase one, or build one instead. When you have the fortune to hire an engineer who is a human being, let him remain a human being.

As an engineer, you are a practitioner; to non-technologists, you also represent technology. In these roles, strive to automate those things that should be automated. Strive earnestly to prevent from being automated those things that should not be automated.

For every program that you write, consider its efficiency and its usefulness. Consider its safety, its impact on privacy, and its effect on the environment. Consider its beauty and its technical merit.

Consider these things and more. Programs, like literature and like architecture, last well beyond their instantiations. In different ways and to different degrees they remain forever as instances of your creativity, of your priorities, and of your decisions.

— Anonymous

This statement has nothing to do with technology. It is a declaration of power, a claim that the rightful place of the programmer is in the center of the universe with everything else revolving around that. This is a political document. It espouses the exemplar of the days of computing when programs were the achievement of an elite priesthood of solitary creative individuals. It claims that power and freedom are a programmer's inalienable right. Customers and managers are welcome to stand by and marvel, checkbook in hand, so long as they keep their uninformed opinions to themselves.

This sense of enfeeblement and powerlessness will be familiar to anyone outside this elite priesthood. It is especially familiar to those such as myself, who scuttle through the no-man's land between the warring factions, trying to get the two sides to communicate.

The programmers hold all the technological aces. Our fingers control the keyboards and only we fully understand what is going on inside the computer. Consumers are powerless even though they are vastly more numerous, and even more importantly, they control the strategic resources upon which the technologists are dependent, the money that pays our salaries.

3.7 PROCESS-CENTRIC VERSUS PRODUCT-CENTRIC

The recently-formed ANSI C++ committee, X3J16, has the task of standardizing the C++ language. Part of this is to specify zero or more standard libraries. Which libraries become part of the standard is still an open question, and will not be settled soon. So for the next year or two, you are on your own.

— Usenet News circa 1990[4]

The tension between the intangibility imperative of the neo-Platonist software priesthood and the tangibility imperative of their neo-Machiavellian consumers, is only one way of characterizing the opposing factions of the software industrial revolution. This section proposes a second way of characterizing the tension in process-centric versus product-centric terms, as these terms are used in Fig. 3.1.

The quotation at the beginning of this section typifies the process-centric focus of today. This task force has set its priorities strictly according to the establishment dogma. The established dogma is that the most productive approach is to concentrate on improving the processes for building software. This relegates a focus on the products of these processes to be dealt with as a secondary priority.

The following quotation, taken from the quarterly publication of a prestigious software engineering consortium, advocates the established process-centric viewpoint quite explicitly.

---

[4] This quotation reflects the attitude of this committee when it originated, not its attitudes today. Library standards have since been undertaken as part of this committee's charter.

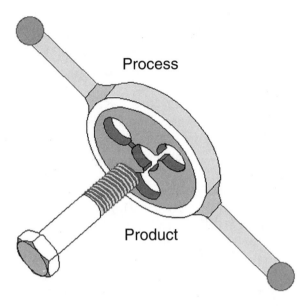

Figure 3.1 Mature engineering domains define standard products, such as this bolt, and encourage diverse processes, such as this die, to be used in making them. In software we do the reverse, defining standard languages and methodologies and expecting standard components to automagically ensue.

Process improvement is central to the ... Consortium's long-term mission to forge significant advances in software engineering. The Consortium believes that the application of new processes or methods to recurrent problems offers significant potential for increased quality and productivity. In this view, process is the integrating "glue" with which distinct methods and tools can be implemented to address the specific needs (i.e., requirements, verification, etc.) of Consortium member companies.

— Quarterly Publication of a prominent research consortium

This process-centric worldview of the intangibility imperative has been the established paradigm ever since I entered this business, the belief that better languages and methodologies, plus greater dedication, brilliance, and freedom for the programmers, will fix what ails us.

The main dissention we hear comes from competing process-centric views. Some advocate one programming language or another. Some prefer Cobol, with or without structured and/or object-oriented extensions. Others prefer Eiffel, or Smalltalk, or C++, or Objective C. Others advocate even higher-level languages based on logic and mathematics. Others prefer analysis and design instead of just coding. Yet others promise rapid prototyping, or CASE tools, or specific analysis and design methodologies.

However, the alternative product-centric viewpoint has been continually present throughout the same period, although continually in distinctly second place. Every programming language, early and modern, provides ways to build and use libraries of well-understood and trusted software components. It has always been technically feasible to tabulate their properties in software engineering catalogs that play the same role as the catalogs of tangible engineering domains.

The opposing faction has managed to get what we call 'software reuse' into the agenda at technical conferences. However, this ideal is well short of being established as a part of the mainstream routine. This group decries the emphasis on processes for fabricating software from first principles. It argues that this approach has brought at best marginal improvements after a quarter-century of pursuit, and that significant gains can only be expected from figuring out how to build a commercially robust market in prefabricated software components so that higher level applications can be assembled from catalogs of trusted parts.

3.8 THE STRUCTURE OF SCIENTIFIC REVOLUTIONS

NEW YORK (DEC. 17) BUSINESS WIRE - "Paradigm shift" enters the business vocabulary as the "hottest buzzword since 'excellence'," according to the current (January) issue of WORKING WOMAN magazine. The magazine credits Joel Barker, a teacher turned management consultant and author of Discovering the Future: The Business of Paradigms, with borrowing the term from science and applying it to business.

—Working Woman, New York

Contrary to this quote, "paradigm shift" is not a term from science. It comes from the history of science. Thomas Kuhn never provided a precise definition for the term 'paradigm',[5] preferring to define it in terms of examples. The synonym that he generally used is 'exemplar', an influential example, analogy, or set of experimental observations that a scientific community adopts as a source of fundamental insight into its world.

---

[5] Thomas Kuhn never used the term 'paradigm shift'. This seems to have originated from his disciples.

Before Kuhn's book, science was believed to proceed incrementally, by steadily adding new knowledge to old. Kuhn argued that understanding actually advances spasmodically. First there is a "crisis" during which earlier worldviews, which he called "paradigms," are recognized as failing. Then there's a revolution in which the old paradigm is destroyed and replaced. Kuhn called these upheavals "Scientific Revolutions," a concept that subsequently became popular by the name "Paradigm Shift" (although Kuhn seems not to have used this particular phrase).

> *The transition from a paradigm in crisis to a new one from which a new tradition of normal science can emerge is far from a cumulative process, one achieved by an articulation or extension of the old paradigm. Rather it is a reconstruction of the field from new fundamentals, a reconstruction that changes some of the field's most elementary theoretical generalizations as well as many of its paradigm methods and applications. During the transition period there will be a large but never complete overlap between the problems that can be solved by the old and by the new paradigm. But there will also be a decisive difference in the modes of solution. When the transition is complete, the profession will have changed its view of the field, its methods, and its goals.*

> *One perceptive historian, viewing a classic case of a science's reorientation by paradigm change, recently described it as "picking up the other end of the stick," a process that involves "handling the same bundle of data as before, but placing them in a new system of relations with one another by giving them a different framework.*
> — Thomas S. Kuhn, *The Structure of Scientific Revolutions;*
> University of Chicago Press; 1962

Of course, Kuhn's contribution was far more than introducing a newly fashionable term. His contribution was to propose the discontinuous model of scientific advancement shown in Fig. 3.2.

Most of the time, practitioners are engaged in what Kuhn calls 'normal' science. Normal science advances paradigms hardly at all. Instead, scientists are engaged in what Kuhn calls 'puzzle solving', exploring the limits of the already well-established paradigm. Kuhn put it this way:

> *Few people who are not actually practitioners of a mature science realize how much mop-up work of this sort a paradigm leaves to be done or quite how fascinating such work can prove in the execution. And these points need to be understood.*

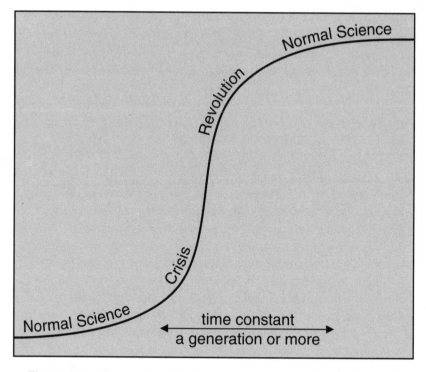

Figure 3.2 Although individuals can undergo paradigm shifts in milliseconds, it can take much longer for an innovation to diffuse through a population.

Mopping-up operations are what engage most scientists throughout their careers. They constitute what I am here calling normal science. Closely examined, whether historically or in the contemporary laboratory, that enterprise seems an attempt to force nature into the preformed and relatively inflexible box that the paradigm supplies.

No part of the aim of normal science is to call forth new sorts of phenomena; indeed those that will not fit the box are often not seen at all. Nor do scientists normally aim to invent new theories, and they are often intolerant of those invented by others. Instead, normal-scientific research is directed to the articulation of those phenomena and theories that the paradigm already supplies.
— Thomas S. Kuhn, *The Structure of Scientific Revolutions*

Eventually this exploration may uncover observations that cannot be explained by the old paradigm. Most of the time these contrary observations can be ignored. It is by no means uncommon for experiments to provide inconsistent or contradictory data. But observations occasionally emerge that can neither be explained nor ignored, thus triggering a crisis that *may* lead to a revolution in which the established paradigm is overthrown and replaced.

Paradigm shift is not synonymous with "drastic, hard-to-grasp change." Paradigms can be quite easy for any particular individual to grasp. The flavor of logical paradox is obvious only when we regard paradigm shifts solely as an individual phenomenon. Paradigm shifts are something that individuals can do in milliseconds.

What can be drastically hard is for them to spread from individuals to entire populations. Once enough individuals make this shift more or less simultaneously (and this can be as long as a generation or more), the paradigm shift becomes a scientific revolution. The German physicist, Max Planck, captured at least part of the reason when he said, "Things never change any faster than it takes the old generation to die off."

The drastic, hard-to-grasp aspect emerges during the diffusion of the innovation stage, when the individual's flash of insight spreads from individuals to populations. This is a realm in which logic is rarely the compelling force, since populations are seldom moved by logic alone. The time course is far more influenced by nonlogical factors such as fashion, vested interest, power, and ever-familiar social forces such as "What will the boss think?" and "What is in it for me?"

Change often spreads from individuals to populations with excruciating slowness. This is not because the new paradigm is hard to grasp. Rather the change may be sufficiently easy to grasp that individuals realize that the implications may not be favorable to their interests.

3.9 SOFTWARE INDUSTRIAL REVOLUTION

> *The thing to do with the future is not to forecast it, but to create it.*
> *The objective of planning should be to design a desirable future and to*
> *invent ways of bringing it about.*
>
> — Russell Ackoff, *Ackoff's Fables*

> *Every great scientific truth goes through three stages: First, people say*
> *it conflicts with the Bible. Next they say it had been discovered before.*
> *Lastly, they say they always believed it.*
>
> — Jean Louis Agassiz (1807-1883)

> *The difficulty is to teach the multitude that something can be both true*
> *and untrue at the same time.*
>
> —Arthur Schopenhauer

"What is the 'best' exemplar for thinking about software? Do we gain the most insight into the hybrid nature of software from disciplines that work primarily in the abstract internal world, such as authors, philosophers, mathematicians, theoretical scientists, and logicians? Or do we learn more from those who work primarily in the concrete external world such as plumbers, carpenters, experimental scientists, and hardware engineers?

This way of posing the question brings out a peculiarity that seems to be true of paradigm shifts in general. They almost invariably involve a chicken versus egg paradox, that makes such either/or questions quite meaningless. The same is true of the software industrial revolution. The only 'right' answer is that software really is a hybrid and that both of its parent exemplars have something to contribute. Software is in many ways like thought itself. But it is also a thing, a commercial property produced by teamwork to be bought and sold in shrinkwrap boxes, just as with the tangible products in a plumbing supply store.

Furthermore, abstract-domain workers such as authors and mathematicians live with everyone else in the concrete domain of everyday experience, and concrete-domain workers certainly think. As in the chicken versus egg paradox, software has a foot in both worlds, but is not precisely of either. Software is unique; neither land nor sea, but a hybrid; a swampy marsh that is hard to explain fully by either analogy because it has characteristics of both parent domains.

I launched this chapter in the confrontational dichotomy typical of technical discussions: right versus wrong, good versus evil, black versus white. But now is the time to dispense with the black versus white coloring box. It is time to acknowledge that a realistic picture of software in particular and the world in general, involves shades of gray.

It is time to balance the discussion by asking, "What could the industrial revolution possibly contribute to something so different from manufactured goods as software?" That is, notwithstanding the newly fashionable rhetoric about object technologies allowing ordinary people to reason about software as we reason about the tangible objects of everyday experience, isn't software entirely unlike the tangible objects in a plumbing supply store? Of what possible use is a grandiose term like software industrial revolution? Isn't this term just so much hype, a far-fetched analogy of no direct use in helping us come to terms with the peculiarities of software?

'Software industrial revolution' does not in any way disagree with this assessment. It is merely a convenient handle, a name for a process of slow evolutionary change with ultimately revolutionary consequences. I coined this term, not to create another bandwagon to hype, but to avoid the hype that is obscuring the substance of another name that is very much in fashion, object technology.

Software is clearly quite different from the tangible goods of the age of manufacturing. It is intangible goods typical of the age of information. It is different from pure information in being a hybrid, in some respects a useful tangible tool to be bought and sold just like hammers and saws in a hardware store. Yet in most other respects it is intangible, ephemeral electronic information of exactly the sort the world is still trying to figure out how to manage as we enter the information age.

As with any hybrid, although software inherits characteristics of both parents, it is precisely like neither. Software is neither land nor sea, but swamp, a confusing and problematic domain. Although the army knows the land, and the navy the sea, neither is quite right for this hybrid abstract/concrete marshland we call software.

3.10 ESSENCE VERSUS ACCIDENTS OF SOFTWARE ENGINEERING

Not only are there no silver bullets now in view, the very nature of software makes it unlikely that there will be any ... no inventions that will do for software productivity, reliability, and simplicity what electronics, transistors, and large-scale integration did for computer hardware. We cannot expect ever to see twofold gains every two years.

First, one must observe that the anomaly is not that software progress is so slow, but that computer hardware progress is so fast. No other technology since civilization began has seen six orders of magnitude in performance-price gain in 30 years. In no other technology can one choose to take the gain in either improved performance or in reduced costs. These gains flow from the transformation of computer manufacture from an assembly industry into a process industry.
— Fred Brooks, No Silver Bullet; Essence and Accidents
in Software Engineering

Now let's return to Brooks' paper with this history of science in mind. But precisely as Brooks cautions, let's not assume that software and hardware automatically have anything in common, apart from the fact that they are both products of the organized activity of people. Rather let's think of today's phenomenal achievements of the computer hardware industry as the culmination of a long and arduous process in how mankind understands and produces tangible things like computers. This process began in the scientific revolution, accelerated during the industrial revolution, and is now operating at peak efficiency within the semiconductor industry.

It took thousands of years to bring hardware engineering to the maturity we see in the hardware catalogs. Software is not at the culmination of this movement, prepared to absorb the process-centric innovations that dominate modern-day manufacturing initiatives. Software is not yet in the twentieth century, but in the Dark Ages. It is still dominated by the Platonic ideal, Ptolemaic cosmologies, alchemists in search of philosophers' stones and techno-centric silver bullets.

Brooks was restating the established paradigm when he said, "These gains flow from the transformation of computer manufacture from an assembly industry into a process industry." Rather, these gains flow from *every* insight into the nature of tangible things of the past several millennia. Although programmers look yearningly at process-centric manufacturing initiatives that are so much in the news in the last decade, the success of the computer industry owes an even larger debt to the product-centric, market-based initiatives that came into prominence during the industrial revolution. If the hardware industry were to ignore assembly initiatives (assembly of interchangeable components) and start building computers and hi-fi sets solely with process initiatives, fabricating every component from first principles in the cut-to-fit fashion of preindustrial revolution craftsmen, their productivity would be no better than that of the software industry today.

The software industry has much to learn from the hardware industry. But we must start the learning process with the old lessons, not the new ones. It is a mistake to believe that we can immediately apply the manufacturing initiatives of the last decade or so without first putting in place the lessons of the preceding millennia. Because of the brief existence of software and its fundamental differences from hardware, software is not yet in a position to directly absorb the process-centric insights of modern-day manufacturing.

Rather, software engineering is comparable to the craft-centered manufacturing techniques before the industrial revolution, and computer science is most comparable to the alchemy of medieval times. Just as Brooks points out in the following quotation, the dominant paradigm in software today is similar to the beliefs of pre-Copernican philosophers like Aristotle and Plato.

> *Second, to see what rates of progress one can expect in software technology, let us examine the difficulties of that technology. Following Aristotle, I divide them into essence, the difficulties inherent in the nature of software, and accidents, those difficulties that today attend its production but are not inherent.*
>
> *The essence of a software entity is a construct of interlocking concepts; data sets; relationships among data items, algorithms, and invocations*

of functions. This essence is abstract in that such a conceptual construct is the same under many different representations. It is nonetheless highly precise and richly detailed.

I believe the hard part of building software to be the specification, design, and testing of this conceptual construct, not the labor of representing it and testing the fidelity of the representation. We still make syntax errors, to be sure; but they are fuzz compared with the conceptual errors in most systems.

If this is true, building software will always be hard. There is inherently no silver bullet.

The computer science establishment believes that the true 'essence' of software can be appreciated only in the abstract world of the mind, and not from the experiential scientists' reliance on experimental observation in the world of the senses.

The product-centric viewpoint of the scientists and engineers has not yet evolved in software, for one extremely good reason. So long as there are only abstract data types and no concrete ones, there is nothing to play the role that nature plays for physical scientists, or that catalogs of standard raw materials play for the engineer. So long as everything we encounter in software is truly new, fabricated from first principles and not assembled from pre-existing parts, how can we possibly hope to benefit from anything the manufacturing age might have to offer?

3.11 THE FOUR OBSTACLES

People are always blaming their circumstances for what they are. I don't believe in circumstances. The people who get on in this world are the people who get up and look for the circumstances they want, and, if they can't find them, make them.
— George Bernard Shaw, *Mrs. Warren's Profession*

Brooks' paper continues by explaining why the four items in this list represent software's inevitable essence. The remainder of his paper examines a number of modern-day technologies and shows why he expects them to bring only marginal improvements.

> *Let us consider the inherent properties of this irreducible essence of modern software systems: complexity, conformity, changeability, and invisibility.*
>
> — Fred Brooks

I want to diverge from Brooks' article at this point. Instead of following his argument further (read the original paper; it's well worth it!), let's ensure that his list encompasses the fundamental differences between the tangible objects of the manufacturing age and the electronic objects of the information age.

One thing that is clearly missing is the single-threaded character of the Von Neumann computer universe. Whereas each particle of the physical universe has its own internal clock, ticking away into infinity, the software components in an application share the global control thread of the computer on which they reside.

This is certainly as fundamental a difference as the four that Brooks listed. This difference is the reason we worry about whether a new component, brought in from outside, uses this ever-finite shared resource correctly and efficiently. When a plumber adds a new feature to the heating system in the basement, he doesn't worry that the clock in the living room might stop working, or that the entire house might crash if the new component is faulty.

Another missing difference is the ease with which additional copies of information-age goods can be replicated and transported from place to place. If a plumber runs out of pipe fittings, he's back to the store to buy more. But when we need a new instance of a stack or a queue, we take it for granted that no new fees will be incurred. This is also a fundamental difference because it explains why the promise of reusable software components has never been thoroughly achieved. Ease of replication undercuts the entire market-based system of human motivation that drove the manufacturing age to the achievements that we take for granted today.

Although there might be additional differences that are also fundamental, adding these two new ones, *single-threadedness* and *ease of duplication*, to Brooks' *complexity, conformity, changeability, and invisibility* will suffice to return to the original question. Does this list truly represent an insurmountable obstacle originating from the inescapable essence of software? Or can the same list be viewed from exactly the opposite perspective, as creating an irresistible incentive for those in search of new ways to compete in the information age?

Figure 3.3 depicts the chicken versus egg dilemma of the software industrial revolution as four obstacles that interlock and support each other like a well-defended infantry position. The labels on the first and third of the obstacles in this figure are the same as those in Fred Brooks' paper. I added the other two to highlight the two differences that Brooks neglected to mention.

The difference in viewpoint between this book and Brooks' article is reflected in the fact that I drew these obstacles as judo fighters, obstacles to be overcome by a sufficiently determined assault instead of insurmountable barriers that cannot be overcome even in principle.

The problem, of course, is that the four obstacles are not independent. They interlock and reinforce each other like the bombers in the flying box formations that Allied aviators flew over Germany in World War II. Although Axis fighters could easily manage to shoot down any individual bomber, the box formation allowed each bomber to reinforce its neighbors and mount a nearly impregnable defense.

Solving interlocking and mutually supporting problems like these is never easy, and no single recipe is infallible. We could easily overcome the complexity obstacle precisely as it was overcome in mature domains like plumbing, by using specialization of labor to distribute complexity so thinly over a software components marketplace that it becomes manageable at each node in the hierarchy.

But what about the supporting obstacles? How, exactly, would such a marketplace function in practice? How could we buy and sell something as intangible as software, something that cannot be measured by the pound or even detected by the natural senses because of the invisibility of Obstacle #3? What would it mean to buy, sell, or own something that is so easy to replicate that it seeps like gas through its owner's fingers? How would we reconcile the difference between tangible goods of the manufacturing age such as silicon chips and information-age properties like Software-ICs[6] that can propagate through networks at the speed of light?

The single-threadedness obstacle derives from the sequential, single-threaded nature of Von Neumann computer architecture. Unlike ordinary objects that operate independently of one another, software objects must share a machine's thread of control. This make them interdependent in ways that ordinary objects are not. If a plumbing component is bad, it is not likely to crash the entire house the way a bad software component can crash an entire application.

Technologies capable of ameliorating this property are well known, even though they're not yet routinely used in practice. Coroutining (lightweight multi-tasking), timesharing (heavyweight multi-processing), distributed computing, and, someday, even concurrent hardware architectures all address the single-threadedness obstacle in different degrees. Exception han-

---

[6] This is not a rhetorical question. It is *the* question upon which everything else in this book depends. Although the software community is not yet ready to consider them seriously, I believe that workable answers have already been explored in television, radio and cable broadcasting. This is why I've deferred most discussion of this most crucial of all of these obstacles until the closing chapters of this book.

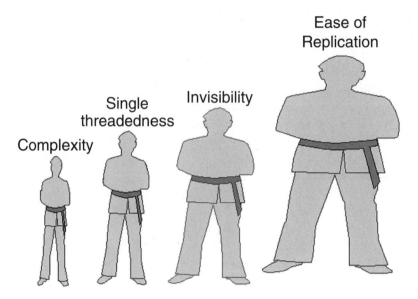

Figure 3.3 represents the difficulties of adopting a more industrial approach to software as a team of four mutually supporting judo fighters. Complexity is the most easily recognized obstacle; it is the one we encounter first. But it is also the most easily overcome, and formidable only because it is supported by the even larger problems in the background.

dling mechanisms relate to the same problem in that they can be used to put an errant thread of control back on track.

The invisibility obstacle could be overcome at least partially with the aid of graphical user interface technology, browsers, and multimedia, and a host of other technologies that are now becoming popular for precisely this reason. And we could also begin doing what real scientists and engineers do when they need to observe something that they can't otherwise see: build tools, such as telescopes, microscopes, measuring instruments, and inspection gauges, for observing what would otherwise be invisible.

I'll discuss this approach in more detail later, in Chapter 5. But what about the supporting obstacles? Most developers can barely afford to build software functionality in the first place, at least partially because of the debilitating effect of the Ease of Duplication obstacle on revenues. Where will the money come from for providing tangible user interfaces for small-granularity software components? And even once robust specification/testing technologies have been provided, where will the revenue come from for applying them in practice?

3.12 ROBUST ECONOMICS FOR INFORMATION-AGE GOODS

The key to this nest of mutually supporting obstacles is Obstacle #4: Ease of Duplication. This obstacle is crucial because, so long as it is intact, it allows no better option than to build software the way Mrs. Kahootie's plumber wants to proceed, by fabricating everything from first principles. Without a sound economic basis for a robust software components market, there is no basis for specialization of labor. The ease of duplication obstacle stands directly in the path of this market by eliminating the financial incentive to provide software components that are sufficiently well tested, packaged, and documented to be any real improvement over what any consumer can fabricate from first principles.

The 'stuff' that we'll have to somehow focus on at the center of this new product-centric world is not a tangible thing of the age of manufacturing, something that can be bought and sold by the copy, like automobiles and cornflakes. Software is intangible stuff that can be replicated as easily as thought itself. If software in particular, and information in general, could be quantified; that is, captured, encapsulated, weighed, and counted, as easily as we weigh cornflakes, we'd have done so already. How can commercially robust information-age markets support the sale of something so intangible that it cannot be captured in order to 'own' it, or quantified to be bought and sold?

Everyone knows software is expensive. This is certainly true of large software objects like spreadsheets and word processors. But it is even more true of small granularity 'reusable' components whose surface area is infinitely larger because they must work correctly in diverse applications. Regardless of the crucible in which these components are formed, be it a components group within a larger organization, an independent commercial entity, or an organization that bundles reusable software components inside a larger software product, what provides the fuel? What is the economic basis for software reuse? Sure, everyone knows reusability is wonderful for consumer side of the reuse transaction. But what makes it wonderful for the producers?

However, even this largest and most formidable obstacle is subject to techno-administrative solutions. There are other industries, older and far more mature than software, that face the same problems with ease of duplication we do. For example, the music recording industry has managed to thrive as radio and TV broadcasting made their product become increasingly intangible during the manufacturing- to information-age transition. The pay-by-use solutions that they pioneered, and which ultimately became their dominant source of revenue, might also be applied to supporting a commercially robust market in software components.

In the case of manufacturing-age examples like Mrs. Kahootie's plumber, the economic basis is pay-per-copy. It works in the souk and it works in the mall. It fuels the economic engine that drives the cooperative specialization of labor hierarchy that makes plumbing so 'simple' compared to programming. The plumber's job is simple because he works at a single level of granularity without having to notice the very real complexity of his task. A finely tuned global economic system has distributed the complexity quite evenly across time and space, to different competing, specialist companies, so thinly that it is hardly noticeable to any. Mrs. Kahootie can concentrate on her home chores and remain oblivious to the finely tuned system of plumbers, retail and wholesale outlets, factories, refineries, and mines, that willingly, even eagerly, compete to help her get dinner to the table.

Pay-per-copy is the primary basis for information-age economics today. Although we buy books and magazines for their information content, no one has ever discovered how to charge for information directly, according to its value to the individual consumer. Since the value of pure information is expressed entirely in the consumer's mind, this value could only be measured directly by mind probes or mental telepathy.

But software has one very great advantage over pure data, knowledge, or wisdom. The value of software is expressed outside the mind, inside a computer where its usage can be monitored as a measure of its value to its consumer. Unlike concrete things and intangible information, using a piece of software involves logic that could report the software usage for billing purposes. Since software is fundamentally unable to monitor its copying but trivially able to monitor its use, this could be a more robust means of measuring the value of software to the consumer for the purposes of commerce than the pay-to-acquire schemes of today.

3.13 SUMMARY

Brooks' article identified a number of ways in which software objects are very different from tangible objects. However, the people who produce and consume software objects are no different from those who produce tangible things. The software industrial revolution involves focusing on this to discover how other people-intensive enterprises addressed complexity in the past. They did it by using commercial incentives to distribute complexity over time and space by using commerce to spread it among other companies who fabricate lower-level components for others to assemble into higher-level solutions.

This chapter has discussed four obstacles that inhibit us from applying the same approach to software that arise from the dissimilarities of software with the tangible objects of everyday experience. I have outlined the technical means by which each obstacle can be overcome, at least partially.

- *Complexity:* Reusing software deals with the first obstacle by distributing complexity over time and space, exactly as specialization of labor reduces the complexity of mundane tasks such as home plumbing.

- *Single-threadedness:* Although multi-tasking, exception handling, and distributed computing are still not widely used, they also show great promise for allowing us to reason about software as we reason about the multi-threaded objects of everyday experience.

- *Invisibility:* Browsers, high-resolution graphics terminals, and text editors are widely accepted technologies for overcoming the inherent invisibility of software. However, we have barely begun doing what scientists and engineers do when they need to study something beyond the grasp of their natural senses. They build instruments, telescopes, and microscopes that make such study possible. Later I'll discuss tools of this nature, specification/testing tools as distinct from implementation tools.

- *Ease of duplication:* This is the most formidable obstacle of all because it subverts the common-sense motivational issues of economics upon which all of the other solutions depend.

This way of ordering the issues amounts to a frontal assault on the mutually supporting obstacles depicted in Fig. 3.3. However, we've never used the jujitsu approach of turning the strengths of an obstacle against itself.

The approach I'll be advocating in this book is one example of turning the very aspect of software that's presently causing us so many problems into being our strongest asset. This is simply to convince the ease of duplication fighter to defect, to switch sides, and work with us to help defeat the other three obstacles. This approach will be detailed in Chapter 6.

The solution, should we ever muster the determination to deploy it, is to craft a better way of inciting others to provide electronic components that actually meet our needs. Once robust incentive structures are deployed, people will discover how to build components other people are prepared to pay for even with the most primitive tools imaginable. And, of course, improved processes and tools will naturally follow.

Chapter 4

Software Architecture

As with most media from which things are built, whether the thing is a cathedral, a bacterium, a sonnet, a fugue or a word processor, architecture dominates material. To understand clay is not to understand the pot. What a pot is all about can be appreciated better by understanding the creators and users of the pot and their need both to inform the material with meaning and to extract meaning from the form.

There is a qualitative difference between the computer as a medium of expression and clay or paper. Like the genetic apparatus of a living cell, the computer can read, write and follow its own markings to levels of self-interpretation whose intellectual limits are still not understood. Hence the task for someone who wants to understand software is not simply to see the pot instead of the clay. It is to see in pots thrown by beginners (for all are beginners in the fledgling profession of computer science) the possibility of the Chinese porcelain and Limoges to come.
— Alan Kay, "Computer Software"; *Scientific American;* September 1984

Objects have become all the rage. There are object-oriented databases and object-oriented drawing programs! There are object-oriented telephone switching systems, object-oriented user interfaces, and object-oriented analysis and design methods!

And oh yes, there are object-oriented programming languages. Lots of them, from Ada at the conservative right to Smalltalk at the radical left, with C++ and Objective-C somewhere in between. And everyone is asking, What could such different technologies possibly have in common? Do they have anything in common? What, if anything, is an object, indeed?

4.1 WHAT, IF ANYTHING, IS AN OBJECT

This chapter is the outcome of my own struggle to understand, and to explain, why this most ordinary of words has been causing so much excitement and hope on the one hand but so much confusion and controversy on the other. It is an attempt to take the question seriously and to answer it fully, without the narrow, ad-hoc, and contradictory definitions so widespread in technical and nontechnical literature of the last decade.

The problem is that trying to define 'object' is like trying to define 'thing'. Such words encompass so much that they elude any attempt to grasp them precisely. Even ordinary adjectives, words that we use everyday without the slightest trace of confusion, 'small' or 'fast' for example, adopt entirely different meanings as the domain shifts among nuclear physics, chemistry, gardening, and geology.

To astronomers, object can mean galaxies one moment and individual suns, planets, moons, and asteroids the next. To biologists, object can mean entire ecosystems one moment and individual plants, animals, organs, tissues, cells, and even molecules the next. To geologists, object can mean entire continental plates, or individual hills and mountains, or the microscopic particles of gneiss, feldspar, and shale in a pebble. To scientists in general, object might mean molecules, or atoms, as easily as the subatomic particles from which atoms are ultimately composed. In computer engineering, object can mean individual gates to a chip designer, chips to a card designer, and cards to a personal computer enthusiast.

Yet in spite of this opportunity for confusion, it never results in the controversies that have raged in the software community for the last decade. The problem is not that this word is poorly defined. The problem is far worse than that. The problem is that *object really means different things to different people*, for perfectly good and valid reasons. Although object means entirely different things to the diner, the baker who made the bread, the farmer who grew the wheat, and the chemical plant who manufactured fertilizer from elemental nitrogen in the air, this causes no problems in well-developed tangible domains. However, it creates nearly insurmountable difficulties in the primitive domain we call software.

In our murky domain of intangible abstractions, it is all too easy to lose our bearings, to misunderstand the context, to confuse the very small with the extremely large. The denizens of the software domain, from the tiniest expression to the largest application, are as intangible as any ghost. Since everything is reinvented from first principles by each individual programmer, everything we encounter there is unique and unfamiliar, composed of components that have never been seen before and will never be seen again and that obey laws that don't generalize to future encounters.

Software is still a place where dreams are planted and nightmares harvested, an abstract, mystical swamp where terrible demons compete with magical panaceas, a world of werewolves and silver bullets. As long as all we can know for certain is the code we ourselves wrote during the last week or so, mystical belief reigns supreme over quantifiable reason. Terms like 'computer science' and 'software engineering' will remain oxymorons—at best, content-free twaddle spawned of wishful thinking and at worst, a cruel and selfish fraud on the consumers who pay our salaries.

The controversies spawned by the strangely hybrid nature of software, like real objects in some ways but quite unlike them in others, have raged from the very beginnings of software. The word 'object' is only a fresh incentive for controversy. Some understand object to mean software that is more *concrete*; composed of tangible objects that can be seen on the screen and touched with a mouse, much like the tangible objects of everyday experience. Others argue that objects are primarily about *abstraction*, the ability to build abstract data types and organize them in conceptual hierarchies through inheritance. To some its value lies in encapsulation, but to others it lies in inheritance. To some it means early binding within the compiler, but to others it means dynamic binding at runtime. To some it means tightly coupled software development technologies like Ada and C++, tools for *fabricating* software from first principles like the silicon fabrication lines of hardware engineering. To others it means loosely coupled tools like Smalltalk and Objective-C, tools for *assembling* software from prefabricated components, like the soldering irons that the silicon foundry customers use.

4.2 HETEREOGENEITY

An astronomer was lecturing on the structure of the universe when he was interrupted by a little old lady in the back, "Professor, your notions are completely far-fetched. Haven't you heard that the world really rests on the back of a giant turtle?"

Amused, he responded, "But what does the turtle stand on?" She replied "Another turtle, of course."

He thought for a second and asked, "So what does the very last turtle stand on?" She snapped, "Young man, you'll not trap me so easily. Everyone knows it's turtles, all the way down."

—Anonymous

An object represents a component of the Smalltalk-80 software system. For example, objects represent numbers, character strings, queues, dictionaries, rectangles, file directories, text editors, programs,

> compilers, computational processes, financial histories, and views
> of information.
>
> —Smalltalk-80: The Language and Its Implementation;
> Goldberg and Robson; Addison-Wesley; 1983.

Suppose that we could resurrect early scientists such as Copernicus, Galileo, and Kepler. Imagine that we flew them all to Silicon Valley, taught them computers and software, and challenged them to do for the software crisis what they did for the astronomy crisis centuries ago. We will leave the culture shock of this experience for science fiction writers to explore. These were intelligent individuals with an unquenchable curiosity about the physical universe, and surely just as curious about computers and software. What questions might they ask in coming to understand this hybrid abstract/ concrete stuff? What methods might they follow? What instruments would they invent, and what vocabulary would they coin, to begin applying the scientific method to software?

In many ways, their new challenge would be easier than the one they faced in antiquity. Questions that took millennia to answer could be conclusively dealt with in the introductory tutorial. "Where does software come from?" and "What is it made of?" are known without any possibility of dispute. Software is a creation of man, not of God or nature. Its fundamental particles are known as bits.

Although the fundamental particles of the software universe are known, "What is it made of?" persists at each higher level of granularity. Is software really "objects all the way down" as the Goldberg and Robson quote says quite explicitly? If so, how are big objects such as applications similar to the smaller components from which they were assembled? Exactly how are tiny components like numbers and character strings similar to huge ones like word processors, spreadsheets, and compilers? In exactly what ways are big objects similar to tiny objects, and in exactly what ways are they different?

Some scientists might inquire into the mechanisms that bind objects together. They might seek to determine if these mechanisms are homogeneous at every level of granularity as in the little old lady's view of the universe. Or are they heterogeneous, just as the welds, glue, bolts, screws, and upholstery stitches that join the parts in an automobile are different from the gravitational, frictional, electromagnetic, and nuclear forces at higher and lower levels of granularity? Do the forces that connect application-sized objects across a network have anything in common with the forces that join tiny objects in C++? Do the tightly coupled mechanisms of C++ have anything in common with the loosely coupled mechanisms of Smalltalk? If so, why do these languages seem so different? If not, why are fundamentally different tools so commonly used to typify object-oriented programming languages?

In other words, precisely what does 'object' mean and how does this meaning change at different levels of architectural granularity? These are the central questions of compositional architecture,[1] the study of how the parts of a software system are bound together to make an organized whole.

4.3 COMPOSITIONAL ARCHITECTURES

God is able to create particles of matter of several sizes and figures ...
and perhaps of different densities and forces, and thereby to vary the
laws of Nature, and make worlds of several sorts in several parts of the
Universe. At least, I see nothing in contradiction in all this.
— Isaac Newton

We find the little old lady's cosmology amusing because we know it is so extravagantly wrong. Upon decomposing any whole into its parts, and those into smaller parts, we gradually encounter parts that are not just superficially, but fundamentally different from the whole. For software to advance beyond the little old lady's worldview, we must relinquish her belief in homogeneous, "turtles, all the way down" software architectures. We must understand the reasons for software's messy but essential diversity and treat it, not as a liability, but as an important and powerful asset.

One important consequence of architectural diversity is that it can hopelessly confuse any attempt to provide crisp definitions for the very terms we use to communicate with each other. What do we mean when we say 'software'? Do we mean something as small as a string compare subroutine, or something as large as a word processor? What does 'programmer' mean? Do we mean only those who build modules in languages like C, C++, or Pascal? Or does a Unix shell programmer qualify? How about a Macintosh user who chooses and assembles off-the-shelf objects like word processors and spreadsheets to build a custom solution to personal computing needs? And of course, what do newly fashionable words like 'object' mean? Do the tightly coupled packages of compile-time type-checked languages like Ada qualify? How about the tightly coupled packages of C++? How about the loosely coupled heavyweight objects of the pipes and filters mechanism of Unix? Of Smalltalk? What about the tangible 'objects' on the glass of a Macintosh screen?

---

[1] Architecture is another of those words with as many different meanings as there are points of view. I will be using it in the sense of compositional architecture, that is, from the viewpoint of how an object is composed from smaller objects.

The problems this architectural diversity can cause are all too familiar to anyone who has witnessed the language wars that have raged with regard to every one of these questions. In the truest black versus white, right versus wrong traditions of software, continuum phenomena cannot be tolerated. Those whose interests and needs draw them to focus on a particular level of this continuum use terms that make sense within that region, but which clash with the way the same terms are used at adjacent levels.

Figure 4.1 is an example of what I mean. A physicist would call this a mass spectrogram of the software I've accumulated on my Macintosh backup tapes over the years. In this particular example, terms like 'software', 'object', and 'programmer' are well-defined. That is, 'object' and 'software' mean large-granularity things that Macintosh users call 'applications'. This figure includes freeware and shareware downloaded from networks such as Usenet, Compuserve, and GEnie plus the commercial software I've purchased over the years. The number of programs of each size is shown on the vertical axis.

Even in this narrowly constrained context, object means at least two different things. Of the 1202 programs in this histogram, the overwhelming majority were small enough (90.6% < 250K) that an individual programmer could build them unaided. This doesn't consider the multitude of smaller programs that never made it into these archives because they were developed solely to be used by their author.

The second meaning can be seen in the labeled points at the right side of this figure. The labels indicate those programs that appear on various best-selling lists. Notice that the labels cluster at the right of this figure. That is, the figure shows that the software that customers pay money to buy correlates quite closely with size. Even in the narrowly constrained domain of this figure, object means two entirely different things. If software is measured democratically, according to the number of programs generated, it is small, abundant, and inexpensive. But if it is measured financially, either by money spent in development or money obtained in revenues, software is large, scarce, and very expensive indeed.

The software crisis is not about the majority of programs, since the majority of programs are small. There is no shortage of small software, nor of individuals capable of writing more. The software crisis is about the huge programs at the right of this figure. These large programs are rare (10% larger than 250K, 4% > 500K, 0.33% > 1000K), expensive, and far more difficult to produce than the small ones.

Software crisis refers to the minority case: commercially important software. Commercially important software is too large to be brought to completion by any single individual and is built almost entirely by organizations. Measured by number of programs written, programmer means a solitary

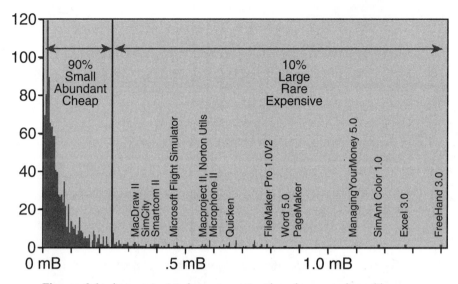

Figure 4.1 demonstrates how program abundance varies with program size within a population of programs consisting of the applications, desk accessories, and control panels in my personal collection of Macintosh programs when I did this study in 1992. The figure depicts the number of applications with each disk footprint (size) on the horizontal axis. The labels point out the widely known and successful programs according to the lists of best-selling software at my disposal. The size and complexity of popular software put them beyond the financial and technical capabilities of an individual programmer. Thus most programs by far (in this sample, at least) are small, abundant, and inexpensive. As a rule they are built by individual programmers. A very small number are large, rare, and, in comparison, expensive. Although they are far more expensive to build, they tend to bring the most revenue to their producers.[2]

creative individual. Measured by the commercial value of programs delivered to market, programmer means a disciplined member of a team.

The dominant paradigm of software development today provides little insight into how to build such software effectively. The dominant paradigm is based on the individual programmer model, which is routinely analogized

---

[2] I have a similar chart that shows how program cost depends on program size. I didn't include it here because it would open an issue that I'll take up elsewhere. The interesting thing is that the graph shows almost no relation at all. The most expensive per mBs are often very small ones, since they are also relatively small. And the cheapest per mBs are generally large programs whose vendors make their money some other way than by selling software (Compuserve and America Online access programs, for example).

to other information-intensive activities such as novel writing, theorem proving, and music composition. Such activities are performed by individuals, not by committees. Regardless of the fundamental differences of software from tangible everyday objects, we have no recourse but to turn to tangible domains of experience such as manufacturing for insight into how large projects can be handled by teams.

4.4 SYSTEM 12

Software's structure is invariably determined by the organizational structure that built it

— Conway's Law

For good software and successful organizations, the inverse is equally true.

— Cox's Corollary

A compositional architecture is a set of interface conventions, encapsulation mechanisms for separating internals from externals. Sole responsibility for the internals can be assigned to one part of the organization. Everybody else needs to understand only the external interface. This lets people of diverse skills and interests collaborate and accomplish more than individuals could do on their own. Just as with Cheops and the pyramids. "We mixed the mortar. They made the bricks. Together, we built the pyramid!"[3]

As large programs go, the ones in Figure 4.1 are all midgets. Airline reservation systems, large in-house information technology programs, defense software, and telephone switching systems can be many times larger and involve hundreds or even thousands of programmers. When organizations of this size and problems of this complexity are considered, single-level architectures are no longer sufficient. That is, the brick and mortar model is elaborated at multiple levels. "We dug the sand. They mixed the cement. Together, we made bricks for the pyramids."

System 12, ITT's distributed telephone switching system, pioneered a radically new approach to telephone switching. There was no central computer whose failure might crash an entire exchange. The system consisted of as many as a thousand small computers, each on a separate card. They were connected by high-speed serial buses that carried both digital and analog (voice) signals.

---

[3] The Great Pyramids of Giza were built of quarried stone, not of bricks and mortar. Software isn't made of bricks and mortar either. I am using bricks and mortar figuratively, not literally.

System 12 was a billion dollar project, with activities ranging from custom VLSI design and fabrication, to custom board layout and design, to designing and building an operating system, and application software capable of handling all telephone functions for an entire city. At least 1200 programmers were involved in software development, scattered all around the globe. Operating system development was in Connecticut, programming language development was in England, configuration maintenance and project management was in Belgium, application software development was in Germany, France, and Italy, and site-specific configuration activities went on in every country who purchased an exchange.

In large undertakings such as this one, architecture is the only hope. The most prominent element of the architecture of System 12 was an early form of object technology. But since this was well before object technologies came into fashion, we never called them 'objects'. We called them Finite Message Machines. A single computer might run many finite message machines simultaneously. These 'objects' communicated with each other, and with objects on other machines, by sending messages to each other over the network.

As the previous description of the globally distributed nature of the project should have made clear, this was not the only architectural level in this system. That is, finite message machines were not the only objects in this system. Connecticut was responsible for building a much larger kind of object called an operating system. This object was the mortar for the application program objects being built in Germany, France, and Italy. At this level of granularity, the finite message machine objects from which these larger objects were composed recedes entirely from view.

Imagine what would have happened if the term 'object' had been fashionable back then. Suppose that the language team in England, working on the compiler for the programming language, CHILL, had started speaking of low-level integers and strings as 'objects' as is now fashionable for languages like C++. Suppose that the very same word became fashionable at the next higher level when speaking of finite message machines. And suppose that the very same word had also become fashionable for speaking of the telephone applications with graphical user interfaces! ITT had quite enough trouble with multiple natural languages, each with different words for the same thing. Imagine the chaos that would have resulted had they also adopted the word 'object' to mean three entirely different things.

The software industry has gotten itself mired in precisely this situation today. The software companies that built the applications in Fig. 4.1 amount to a loosely coupled distributed organization exactly like the distributed organization that ITT used to build System 12. Some parts of this organization speak of applications with graphical user interfaces as objects. Others speak of the elemental particles within programming languages like C++ as objects. Yet others speak of the molecular-sized particles within languages like Smalltalk as objects. Is it any surprise that everyone is asking, "What does object really mean?"

The confusion is not that object is ill-defined. The problem is far worse than that. This word simply has different meanings, just as component means totally different things in tangible domains such as astronomy or silicon engineering. But this is a very grave problem indeed. For although mankind has had millennia of experience with the architecture of tangible things, we have less than fifty years of experience with computer software. This is simply not enough time to develop a broad consensus for architectural levels and a vocabulary for discussing them in juxtaposition. Whereas terms like astronomy, geology, chemistry, and nuclear physics are readily understood to refer to objects of radically different sizes, there has been insufficient time to identify and name the analogous architectural levels in software, whether object-oriented or not.

4.5 FABRICATION AND ASSEMBLY

From the inception of software, we've thought of programming as analogous to abstract activities like authoring a book, composing a poem, or proving a mathematical theorem. The very words that we use in speaking about software evolved from this view that software is a kind of information, something of the internal world of the mind, more akin to *thought* than *thing*. We create software as authors create novels, by 'writing' it in a programming language. Like mathematicians, we strive to 'prove' it correct, but certainly not by assembling it from prefabricated parts the way everyday objects like automobiles and airplanes are built.

It is not surprising that this solitary, creative, free spirit model of programming is so dominant. Mankind's experience with organizing to solve large problems goes back a very long way. It is hardly surprising that we've acquired a substantial vocabulary for talking about the activities at various levels of the specialization of labor hierarchy. Even a mundane activity like plumbing a new house involves a labor hierarchy of thousands of individuals, distributed all around the globe. Some specialize in mining and others in refining. Yet others fabricate elemental components such as faucet components and pipes from raw materials. Others specialize in assembling these prefab components into higher-level components, and so forth until the assembly process culminates with the plumber.

The same distinctions apply to software, except that software begins where the hardware engineers leave off. That is, the mining and refining have been done by others, so that software is concerned entirely with fabrication and assembly. There is no sharp dividing line between them for they are only regions of a continuum of architectural possibilities. The terms distinguish regions within a continuum distinguished by such continuously varying quantities as reversibility, amount of specialized skill needed to operate at that level, and so forth.

The name of this bundle of continuously varying quantities is encapsulation. Encapsulation is not a binary either/or quantity. It can vary continuously from tight to loose. At the fabrication end of the encapsulation continuum, encapsulation is extremely loose. To say the same thing another way, the coupling between components is extremely tight. Although it takes highly specialized training and tools to work in a tightly coupled fabrication environment, the tight coupling leads to performance and other advantages that cannot be achieved in a loosely coupled, tightly encapsulated assembly-style environment.

For example, silicon chips are produced using fabrication-style components. Small armies of scientists and engineers can struggle for years to produce lithography masks for making a new CPU chip, before turning them over to manufacturing. Such fabrication processes are irreversible—a chip cannot be readily converted back into clean wafers.

It certainly requires far more skill to design and build a silicon chip than it does to use assembly-level technologies. Anyone can assemble a home computer from purchased components, or even to build one with intermediate technologies such as prefabricated chips and discrete components.

At the *assembly* end of this continuum, encapsulation is extremely tight. Conversely, coupling is extremely loose. The tight encapsulation helps to hide the complexity of the internals of a product so that the product can be used by the broadest possible constituency. Building cards from chips and discrete components is a low-level assembly operation because the cards can be unsoldered to recover the original components. Installing a home computer by cabling together computers, disk drives, and printers is a high-level assembly operation.

4.6 SPECIALIZATION OF LABOR

The software industrial revolution signifies a broader socioeconomic perspective from which to consider why object technology means so many different things to different people. Throughout this book, I'll provide a number of ways of refining the meaning of this term in terms of paradigm shift, such as the process versus product paradigm shift to be discussed in the following section. However, the core meaning of this term is best understood in terms of organizational change, as distinct from a change in technology.

This change is called specialization of labor in manufacturing circles. The change has definite implications on technological matters such as software architecture, and I'll return to those momentarily. But first I want to expose the core of the issue, which is a change in how human beings organize to build complex software systems.

Figure 4.2 represents the established paradigm up until the present day. That is, programming is thought of as an activity of a solitary individual, who creates software much as novelists write books. Signs that this is the established paradigm can be found throughout the very vocabulary we use in speaking of software development. We 'write' software in a programming 'language'; we don't 'assemble' it from prefabricated parts. Figure 4.1 shows that the solitary developer model is by far the dominant paradigm when dominance is measured by the number of programs written.

Figure 4.3 represents the established paradigm for programs that exceed the capacity of the solitary developer model. This is the dominant paradigm when dominance is measured in monetary terms (money spent on develop-

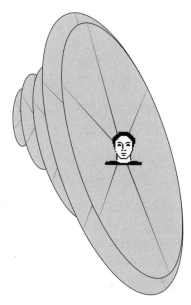

Figure 4.2 is the dominant paradigm for building software to this day, if dominance is measured in terms of number of programs. The organizational structure is the same as that used for producing technical articles, romance novels, or popular songs. Programming is a form of writing, organized as a solitary activity of a single creative individual. The spirals represent iterative development as in the software life-spiral model. The subdivisions within the cycles represent life-cycle stages such as analysis, design, implementation, and testing.

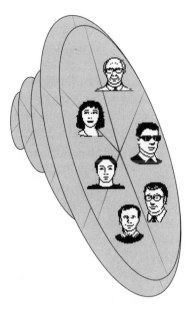

Figure 4.3 is the dominant paradigm when dominance is measured in monetary terms such as money spent on development or money earned in revenue. Although the complexity of the project can be distributed over the team members, the entire complexity is the responsibility of a single team. That is, complexity is not distributed across space and time as it could be with a vertically specialized labor hierarchy.

ment or money earned in revenue), such as with the large programs shown at the right of Fig. 4.1. Instead of having a solitary programmer be responsible for all aspects of a project, we use a software development team.

There are two dominant models for planning such projects. Since one is a subset of the other, both are shown in the same figure. The spiraling pattern represents the currently popular software life-spiral model, where development proceeds from rapid-prototyping to development to maintenance in a number of iterative cycles. The topmost cycle of the spiral is the ever-popular software life-cycle model, also called the 'waterfall model',[4] where development within each cycle proceeds through distinct phases such as analysis, design, implementation, and testing.

Figure 4.4 represents the new paradigm that we'll be exploring in this book. The production of each component proceeds through the usual life-cycle and life-spiral phases. However, the phases used in producing the components are decoupled, through encapsulation, from the phases of the

---

[4]I have always been amused by how apt this term actually is. Isn't a waterfall something that starts out clean and smooth and predictable at the top, and explodes into chaotic turbulence at the bottom?

Figure 4.4 represents the software development paradigm that the term, software industrial revolution, intends to suggest. Specialization of labor distributes the complexity of software development across time and space by allowing each individual to specialize on a single vertically specialized product, such as the reusable software components implied by the hardware components at the right of this figure. The critical difference between this figure and the preceding ones is that software development is now centered on the arrows, not on the nodes. Development is organized around the products that move between individuals, not on the processes that the individuals use to build them. The new paradigm has introduced product-centric innovations such as standards and interchangeability that were absent in the two preceding figures.

overall project. That is, the components can be, and generally would be, produced independently of the projects that use them. In the limit, each node becomes an entirely independent commercial entity. The arrows would then represent a commercially robust market in standardized interchangeable software components at various levels of granularity.[5]

---

[5] This paragraph implies, without pursuing in detail, the very issue that makes the software industrial revolution seem so impossibly remote today. I want to defer the question of how one might buy and sell easily copied commodities like software until we've covered the more familiar technical issues. I will deal with the economic issues in Chapter 6.

This model can be thought of as an evolutionary extension to the model shown in Fig. 4.2. After all, a two-level specialization of labor hierarchy already exists whenever programmers build programs to be used by others. And more ambitiously, the figure suggests that, instead of working for a software development company, individuals might start a home business around being the world's best supplier of some software component. In other words, Fig. 4.4 can be viewed as returning to the days when programming was a solitary activity of a single creative individual.

The new model can also be thought of as an evolutionary extension to the programming-by-committee model that most large companies use today. That is, the software development organization shown in Fig. 4.3 might be further specialized into internal component development groups that produce reusable software components that other groups in the same company will use.

Although the software industrial revolution will certainly unfold in such an evolutionary manner, I encourage you to recognize that this is really a revolutionary change, not an evolutionary one. Something entirely new has appeared in Fig. 4.4, something that was entirely absent in both of the preceding figures. One name for the difference is standards. Another is interchangeability. Yet another is architecture as I've used that term in this chapter.

A third difference, which the next section will discuss in more detail, is that software development is no longer process-centric, centered on programming languages and methodologies. It is now product-centered, revolving around the production and consumption of encapsulated reusable software components. This is a very large difference indeed, for these smaller granularity software 'products' are unlike the products in, say, a shopping mall. Whereas most products are tangible goods, assets that can be bought and sold by the copy, these smaller granularity software products are electronic hybrids, not quite as ephemeral as thought itself, but completely invisible to the natural senses. They are made of bits, not the atoms of commerce since aliquity.

4.7 PROCESS VERSUS PRODUCT

The ambiguity about what to call objects at various levels of granularity is a symptom of a much deeper problem, one that is uniquely peculiar to software and a sure sign of the youth and immaturity of this domain. We assign names to the *processes* for building and using software, but we fail to define, standardize, and name the *products* of these processes.

Without process-independent names for the *products* of these processes, we find ourselves in the situation of Thag, the caveman who invented the first stone axe and unwisely named it ThatWhichThagProduces. Poor Thag is no longer with us. He was wiped out in a language war with Thor who named the *second* stone axe ThatWhichThorProduces. Things finally settled down when some stone age linguist realized that coining a process-independent term, 'axe', might help to stem the bloodshed.

Smalltalk, Objective-C, C++, and Eiffel are names of programming languages. Programming languages are processes for producing software objects at the architectural levels where programmers spend their professional lives. Names like MS/DOS, Macintosh, Windows, and OS/2 signify operating systems, which are also processes for building software objects at the higher architectural level where end-users spend most of their time.

But surely the software industry is beyond the caveman stage. Tell me, then, what should we call ThatWhichSmalltalkProduces? If you say 'objects', how then shall Smalltalk programmers avoid language wars with those who believe that 'object' means ThatWhichC++Produces? What should we call ThatWhichUsersRun in environments such as Macintosh, Amiga, MS/DOS, and Windows? If the answer is 'program' or even 'application', how do we avoid getting into exactly the same kind of wars with those who use time-sharing systems like Unix, VMS, or MVS where these words mean something similar, but in detail, something entirely different?

Figure 3.1 brings the stone axe example up to date with the part of the industrial revolution in which the advantages of interchangeable parts were discovered. As I will show in greater detail in Chapter 5, the achievements of this period were based on two major innovations. One innovation had to do with *process*, and the other with *product*. The process innovation was the dominance of tools of certainty, such as the die in this figure, over tools of risk that cottage industry craftsmen used, filing screw threads by hand.

The progression from tools of risk to tools of certainty is actually underway in software already. We welcome process improvements such as the replacement of assembly language with higher-level languages up through object-oriented languages and beyond. But the industrial revolution also involved product-centric improvements, and these have still not been adopted in software. It brought interchangeability and standardization to the manufacturing process, as well as powerful measurement and inspection tools for gauging compliance of a component to its specification. Our neglect of product-centric initiatives can be seen not only in the absence of analogous software specification/testing technologies; it is particularly obvious in our failure even to assign process-independent names to small granularity software components. Why name components so long as there are no standards and every component is built in a cut to fit fashion?

We are perpetually engaged in warfare among devotees of different languages and unable to adopt the common sense wisdom that the phrase, 'Use the right tool for the job,' conveys in domains far more mature than ours. Look at Fig. 3.1 again with this phrase in mind, and its meaning turns upside down. 'Use the right tool for the job' says that the dominant piece in this figure is not the die, but the bolt. The die is only a process, and it is replaceable according to the task at hand. The die is only one of a large number of perfectly acceptable ways of creating a standard bolt. Each is perfectly acceptable, so long as it meets the absolute requirement of producing a standard bolt. Filing the threads by hand might be preferable for making a one-of-a-kind bolt than building a custom die. Mass production thread-cutting machinery would be a much better choice for long production runs.

Our neglect of product-centric initiatives in software is quite understandable. Software products are intangible, so we focus on the one thing about software we can observe with our natural senses, the tools and activities of producing it. But by not confronting these difficulties and addressing them, we're unable to collaborate effectively. The software crisis is a result.

So which is more important, process or product? Which is the center, the earth or the sun? Which came first, the chicken or the egg? Such questions seem silly when put in this either-or manner, because clearly both are important. But major paradigm shifts throughout history are invariably about precisely such figure/ground distinctions. The earth didn't become less important when Copernicus put the sun at the axis of his calendar computations. Nor will process-centric issues be any less important when we center the software development process around a market in prefabricated components.

4.8 ENCAPSULATION AND THE LAW OF PROXIMITY

I recently watched a wonderful Public Broadcasting System television program about the history of personal computing. The program interviewed Doug Englebart, Ted Nelson, and Alan Kay who pioneered many of the conceptual foundations. Steve Jobs, Steve Wozniak, and others provided the story of personal computer hardware. Bill Gates and Mitch Kapor spoke of the role that companies like Microsoft and Lotus, whose only product is software, played in fueling the phenomenal growth of this young industry.

The discussion turned to differences between the products of hardware and software companies. The program claimed that physical systems like computer hardware, airplanes, and Twinkies are governed by physical laws, but that physical law does not apply to software. The particular law mentioned was the law of proximity. When a hardware system fails, the cause is likely to lie in the topological vicinity of the symptom. In software, failures can arise from defects anywhere in the system.

The very term, physical, implies an object with tangible, definite boundaries so that external influences can enter only through interfaces put there by the designer of the object. Mitch Kapor argued that with software, there are no such encapsulation boundaries and unplanned influences can originate anywhere in the system. The origins of this well-known failure of physical laws for software, the law of proximity in particular, arise from a more fundamental cause: lack of encapsulation in conventional tightly coupled software.

Although this black-versus-white distinction between hardware and software was fine for public television, upon close examination it fragments into shades of gray. Even in hardware there are architectural levels that are barely more encapsulated than software. For example, the designers of state-of-the-art VLSI circuits today see their problems as analogous to those of software. In addition to all of the problems arising from the lack of encapsulation within the tightly coupled design of a complex VLSI chip, they face the additional problems of working so close to the physical limits of their silicon-based medium. And even within this hardware domain, there are even lower architectural levels where encapsulation is almost entirely absent. For example, in manufacturing the raw materials of chip manufacturing, such as the large single crystals of silicon, called boules, from which silicon wafers are sliced, the properties of the boule are amorphously determined by each and every ingredient.

Even in software there are architectural levels of granularity that are encapsulated nearly as strongly as hardware. When a program crashes on a timesharing system such as Unix, diagnosis proceeds much as in hardware, according to topological proximity. The cause is most likely inside the program that crashed because the encapsulation mechanisms within a program are not as robust as those around the program as whole. The problem is not likely to arise in some program my neighbor might be running who is sharing the same machine. Timesharing operating systems provide an encapsulation boundary around the software objects of the level of granularity we call a 'program' that is of considerable help in controlling how widely undesired influences might spread.

The advent of object-oriented programming languages has also made major contributions to enabling the law of proximity for software. Although these languages also support encapsulation, it is for objects several levels of granularity smaller and which are several orders of magnitude more tightly coupled than time-sharing system processes.

4.9 PROGRAMMING LANGUAGES AND OPERATING SYSTEMS

Although our cultural beliefs in the uniqueness of the software have caused us to avoid manufacturing terminology in connection with software, we actually do make the very same distinctions. The traditional fabrication technologies for software are programming languages like C and Pascal, which irreversibly transform a raw material (source code) into something altogether different (binary code). The traditional assembly technologies for software are operating systems like Unix and Macintosh. They let users invoke different programs in combination to solve user-specific problems, exactly as plumbers assemble prefabricated components that they buy[6] from a plumbing supply store.

Back when software was simple, when programmers were primarily concerned with programs like those at the left of Fig. 4.1, it was sufficient to distinguish only two regions on the fabrication-to-assembly continuum. On the one hand, there were programmers, who use programming languages to fabricate programs from first principles. On the other hand, there are users, who use operating systems to assemble prefabricated programs into custom solutions to their own special needs. But as programs grew larger and more complex, new technologies for encapsulating components have been added in which the old either-or distinctions between fabrication and assembly, programmers and users, programming languages and operating systems, are no longer sufficient.

Consider Smalltalk, for example. Is Smalltalk a fabrication technology like C? Or is it an assembly technology like the Unix shell? Choosing between these two pigeonholes is not easy because Smalltalk has characteristics of both. In fact, Smalltalk originated out of the belief that the programming language versus operating system distinction was unnecessary and undesirable in a personal computing environment. Smalltalk is *like* C in that both are used for programming. But the same can be said of the Unix shell. Smalltalk is *like* the Unix shell in that both are oriented toward assembling prefabricated components. But the same can be said of the subroutine linkage and macroexpansion capabilities of C.

One sign of the orientation of Smalltalk toward assembly can be found in the syntax of the language itself. Fabrication technologies such as C provide incredible syntactic diversity, with distinctly different syntax for declaring

---

[6] A huge issue lurks right here that I want to defer until I can deal with it thoroughly. How does one 'buy' and 'sell' intangible property such as software components? If you can't count them or weigh them, what does it mean to 'own' them, particularly since they can be copied as easily as thought itself?

variables from that used for statements and expressions. Each control flow construct has its own special syntax, with 'if' statements, 'while' statements, 'for' statements, and 'goto' statements, each written entirely differently. Smalltalk has none of this diversity. But when you try to pin down the syntax of Smalltalk, poof! It vanishes before your eyes. There's no syntax there to compare. Smalltalk syntax is nothing more than a way of sending messages to objects. Even the syntax for writing control flow constructs such as looping and conditions is nothing more than sending messages to objects of class Boolean!

If you really bear down on the fabrication versus assembly distinction, a similar ambiguity emerges even in thoroughly traditional languages like C. Parts of C are concerned with assembly of prefabricated components from subroutine libraries. Other parts are concerned with fabrication of code from first principles; i.e., the syntactic components that C provides for declaring data and writing expressions and statements.

All this is a sign that software architecture has been advancing very quickly and aggressively to cope with the complexities of the situation shown in Fig. 4.1. But the architectural diversity has outpaced our vocabulary for discussing what is going on. So long as we allow ourselves only two pigeonholes for categorizing new innovations such as Smalltalk, is there any wonder that we keep making meaningless comparisons between very different things forced into the same pigeonholes—for example, by comparing languages as different as Smalltalk and C++, or operating systems as different as Macintosh and Unix?

4.10 A RICHER SET OF ARCHITECTURAL DISTINCTIONS

The original intention of the phrase 'object-oriented' was to signify orienting on the *objects*, not on the processes (languages and methodologies) for building them. However, there is a variety of entirely different meanings of 'objectness' out there today. There are the kinds of objects that programmers have in mind when we speak of object-orientedness in connection with languages like C++, and there are the entirely different kinds of objects that end-users, salesmen, and marketeers have in mind when they speak of applications with graphical user interfaces as being object-oriented.

So which meaning is right? The usual way authors deal with this problem is to bless the meaning held by the audience at hand and to damn the other meanings as hype. Although this has the very great advantage of simplifying the discussion and making both the writer and the reader's job easier, this has always seemed counterproductive to me. Aren't the graphical objects that the end-users and marketeers have in mind *more* like tangible everyday objects than the abstract data types the technical community has in mind? If the controversy over whose meaning is 'right' ever came to a shoot-out, mightn't the technical community *lose*? And for that matter, *shouldn't* their meaning lose? After all, the salesmen and marketeers are merely responding to the demands of the users of software, who are ultimately the ones who pay the programmers' salaries.

Instead of engaging in such right versus wrong debates in this book, I want to follow the more difficult path outlined in Fig. 4.5. I have not chosen this path because it is difficult nor merely to be different from other authors. I have chosen it because the cooperative specialization of labor that this fig-

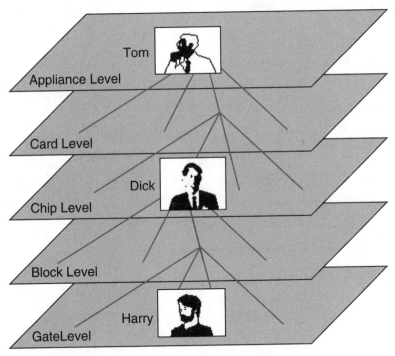

Figure 4.5 shows the compositional architecture of the personal computer hardware industry. The architecture provides at least five different ways whereby electronic functionality can be packaged, ranging from tightly coupled fabrication at the bottom to loosely coupled assembly at the top. Which technique a particular designer can choose is tightly constrained, not only by technical needs of his project, but by whether his client at the next level higher has the skills and motivation to work the various kinds of objects. For example, Tom is not a likely customer for functionality that is packaged as chip- or lower-level objects.

ure implies is the only way to ever bring the software crisis under control. This cooperation cannot possibly emerge so long as core words in our vocabulary, the very words that we use to communicate with groups at other levels of the hierarchy, are being interpreted in today's hard-edged, black versus white, right versus wrong manner.

The figure depicts Tom, Dick, and Harry as stereotypical examples of the vastly different clienteles who gravitate to each architectural level. It seems better to assume each of these groups holds a piece of the truth, but that none of them owns it completely. End-users like Tom reside at the extreme assembly end of the continuum at the top of this figure, where tightly encapsulated, loosely coupled objects are the most common. When end-users, salesmen, and marketeers are communicating horizontally, with others like themselves within the upper levels of this continuum, why should programmers care when they use the word 'object' to mean loosely coupled, tightly encapsulated objects; i.e., programs with graphical, direct manipulation user interfaces?

Programmers such as Harry reside at the extreme fabrication end of the continuum at the bottom of this figure, where weakly encapsulated, tightly coupled objects predominate. When programmers are communicating horizontally, with other programmers at the lower levels of this continuum, why should end-users care that we use the word 'object' to mean the tightly coupled, loosely encapsulated objects of languages like Ada, C++, or even C?

However, when we're communicating *vertically*, with groups above or below us in the specialization of labor hierarchy, we need to be aware that object means entirely different things to constituents of different architectural levels. That is, we need to choose words that generate more light than heat. In particular, we need words that specify exactly what kind of 'object' we're referring to. A plausible set of such words exists in hardware engineering.

Process- and Task-level Objects: The highest architectural level most personal computer users encounter consists of large-granularity appliance-style objects such as ready-to-use computers, printers, modems, and the like. Even ordinary people, those with no specialized hardware knowledge at all, have enough skill to connect appliance-level objects to build a customized solution to their personal computing needs.

Task-level Objects are the next architectural level down. These require slightly more expertise to use correctly, since tasks must be plugged into a bus inside of an appliance. Fewer end-users have the skill and confidence to participate in this market, since installation involves violating the black-box encapsulation (and occasionally the warranty) of the task-level object.

Chip-level Objects: Those with the skill to work with silicon chips occupy a uniquely hybrid niche in the hardware specialization of labor hierarchy. Chip users need specialized electronics training. But for their own special reasons, they've chosen to occupy a hybrid niche in the specialization of labor hierarchy, building task-level objects by assembling them from chip-level objects that were prefabricated further down the hierarchy.

Gate- and Block-level Objects: Those who work with gate- and block-level objects are the highly skilled minority with specialized jobs within the silicon fabrication companies. High skill levels are mandatory, because gate- and block-level technologies are tightly coupled activities, where optimization of the product is far more important than the productivity of the individual designer.

The point of this figure is that object diversity is inextricably connected with people diversity. 'Fixing' the latter is not an option. People as diverse as Tom, Dick, and Harry not only exist in this world, but also coexist within each software development project. Tom, after all, represents the customer, without which few software projects (or programmers) would ever exist.

Since the software industry has never formally recognized that diverse software architectural levels are a sign of maturity and not weakness, we've never agreed on generic names for the objects at different levels, except for names that are narrowly dependent on specific languages and operating systems. So whenever the clientele of different levels meet, time that could be spent learning from one another is wasted in interminable disputes over what the words mean. "Does Ada or C++ support 'real' objects?" "Does the Macintosh support 'real' multi-tasking?"

4.11 FIVE ARCHITECTURAL LEVELS BY EXAMPLE

To concentrate on software architecture for a domain as intangible as software, where the parts lack concrete substance that can be observed with the natural senses, we'll have to rely on some kind of language-independent vocabulary, a set of names that these diverse communities can use to discuss their specialized role in the software hierarchy. Since no process-independent terms have yet been agreed on for software engineering objects, I'll simply adopt the terms that hardware engineers use. This is not to suggest that software components are 'like' hardware components in any useful way. It is only to introduce a set of labels to which precise technical definitions will be assigned later.

It is important to realize that the five architectural levels are not new. They're certainly not something that I invented for this book. I have only assigned names to something that has existed in every environment I've ever used. Nor do these levels exist only in some environments and not others. They are present, either potentially or actually, in every programming environment that has ever existed. What varies is the extent to which each level is *supported* by the environment. The levels are present potentially if not actually, regardless of whether environment provides ready-to-use modularity and binding mechanisms. Programmers can almost always turn this potential into actuality by merely adding the missing support on their own.

Figure 4.6 may help with the subtle but crucial distinctions among these five levels. The figure represents the extent to which the five architectural levels were supported on the most primitive programming environment I ever used. This was an early PDP 8/I computer with 8000 words of 12 bit memory. Its only I/O devices were console lights and switches, an ASR Model 33 Teletype, and a bank of analog-to-digital converters that I used to monitor my neurophysiology experiments in graduate school.

The vertical axis shows the five levels, implying their *potential* presence even in this primitive environment. The horizontal bars show the extent (on a subjective 0 to 10 scale, with 10 representing unattainable perfection) to which each of the five levels was *actually* supported. For example, the figure grants even this most primitive of environments level 1 support for gate-level objects. This support amounted to nothing more than the console switches through which one could enter the state-of-the-art gate-level 'ob-

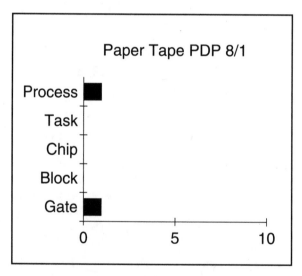

Figure 4.6 Even a primitive PDP 8/I machine with 8K memory, whose only I/O devices were a teleype paper tape reader and console switches, the five architectural levels can be identified. However, they were supported in only the most rudimentary sense since programs were small and few users needed more than one level of granularity.

jects' of that day, raw machine instructions. The figure also grants level 1 support for process-level 'objects', or applications. This support amounted to nothing more than the sheet of paper I kept taped to the console. It held a printout of the machine instructions of a bootstrap loader that I loaded via the switches to load larger programs via the Teletype's paper tape reader.

Three other architectural levels were present even in this most primitive of environments, but they were not supported in any meaningful way. The figure shows no support for block-level objects, or subroutines. These were well known even then and early computers had adequate instructions for defining and using them. The figure depicts the subroutine level as not supported since subroutine-based modularity was irrelevant on an 8K paper tape machine.

This was even more true of chip-level objects (Software-ICs). This is a level that few had even heard of back then. The same goes for task-level objects, or coroutines. Paradoxically, these were as well known as subroutines even in those early days. In fact, the section of the PDP 8/I manual that explained how to build and use subroutines also explained how to build and use coroutines. Since our neurophysiology experiments involved controlling equipment in real time, we used coroutines heavily in our work. The figure shows them as unsupported because we had to build them ourselves, without assistance from the PDP 8/I environment.

Figure 4.7 rolls the calendar forward several years. This figure shows the extent to which the same five levels were supported on the large (for those days) PDP11/70 that we used to support large software development projects when AT&T first released Unix to commercial users. This environment is much richer and it consists of many pieces: the PDP11/70 hardware, the Unix shell, C compiler, the Unix kernel (operating system), and so forth. I could have composed a separate chart to show the contribution each of these pieces provided for each architectural level separately. But for brevity, I've summed the separate charts to produce a composite picture of this environment as a whole.

The chart shows that much has changed during this decade. For example, the bottom row of this chart shows gate-level objects (expression-based integration) are supported far more thoroughly than before. Machine language and console switches have now been supplanted by C compilers which have become fully capable of transforming source expressions into machine instructions. Whereas once we thought of raw machine instructions as the indivisible atoms of software, we now thought of source-level C expressions in the very same way.

The second row shows that block-level objects (subroutine-based integration) are now being supported quite thoroughly. Programs are becoming sufficiently complex that higher-level modularity mechanisms are relevant.

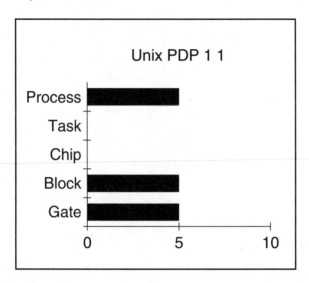

Figure 4.7 The environment shown in this figure was a relatively large PDP 11/70 minicomputer running one of the first commercial releases of Unix (System 7.0). Higher-level languages such as C have nearly obsoleted machine language for gate-level programming, and block-level programming is now thoroughly supported with powerful linkers, librarians, and loaders. Process-level programming is also thoroughly supported with high-level languages such as the Unix shell.

Block-level objects are now supported in every competitive programming environment, as in the subroutine mechanisms of the C compiler and the Unix linker.

Last, but not least, process-level objects (application-based integration) are now supported quite thoroughly. The top row shows that programmers now build applications and store them on a disk so that others can use them later. Even more significantly, timesharing allows all of this to go on simultaneously right on the very same machine. Each appliance-level object runs in its own protected address space, independently of others that might be sharing the same machine. If one user's objects become damaged (and under Unix, each user generally uses several of them simultaneously), no harm is done to other users.

Figure 4.8 rolls the calendar forward another decade to show what Smalltalk brought to the table. The personal computer revolution triggered the emergence of an entirely new level of the specialization of labor hierarchy, as typified by Dick in Fig. 4.5. The interests of this group were entirely unlike those timesharing users such as Tom, to whom precanned but inflex-

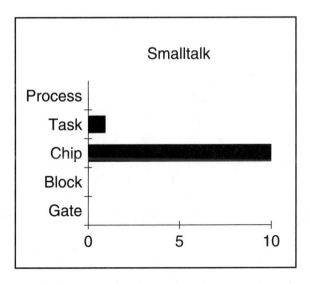

Figure 4.8 Smalltalk represents a radical departure from the timesharing philosophy of Unix and the tightly coupled philosophy of C. Smalltalk grew out of the belief that the traditional distinction between programming languages and operating systems was a reflection of timesharing, which had no place in a personal computing environment. Thus Smalltalk supports neither the fine-granularity, tightly coupled objects of C nor the coarse-granularity, loosely coupled objects of the Unix shell. Its primary focus is on providing a pure programming environment based entirely on an entirely new kind of intermediate-granularity object that my previous book called a 'Software-IC'.

ible Unix-style objects were generally sufficient. Nor were they similar to programmers like Harry who were comfortable with the low-level programming style of C.

Smalltalk was a reaction against the need to have any boundary between programming languages and operating systems. The designers of Smalltalk banished conventional programming by fabrication altogether. Fabrication is now the sole responsibility of a very few programmers, originally at Xerox PARC (Palo Alto Research Center) and today at Digitalk or ParcPlace Systems. These experts are responsible for maintaining the primitive data types inside the Smalltalk virtual machine. All other programmers assemble software by connecting these prefabricated Software-ICs together.

This figure also shows that Smalltalk provides some support for objects of an even higher level of granularity. This support is very meager, so much so that it can be, and usually is, overlooked. It consists entirely of a single class,

Task, whose instances manage an 'independent' thread of control. Just that, a single class. The only mechanisms Smalltalk provides for manipulating these instances is through the general message-passing and browsing capabilities available for any object. Just as the console switches of PDP11 provided level 1 support for gate-level objects, this barren little Task class provides level 1 support for a higher-level kind of object that I've dubbed 'task-level objects'. From this tiniest of seeds, an entirely new level of modularity eventually sprouted in personal computing environments such as the Macintosh. 'Object' has never meant the chip-level objects of Smalltalk in these environments. It means application-sized objects that coexist inside a shared machine address space, exactly as in the Smalltalk Task class.

Tasks, or lightweight coroutine-based processes, are task-level objects. Although task-based objects do not support inheritance, they are even more object-oriented, more like everyday tangible objects, than the chip-level objects from which they were composed. Task-level objects encapsulate far *more* than chip-level objects, and the encapsulation is even stronger. They encapsulate more because a task-level object encapsulates the object execution history (stack arena), in addition to the state (instance variables), and behavior (methods) of chip-level objects. And they encapsulate more strongly since task-level objects are generally packaged inside a graphical user interface that hides the textual interface through which lower-level objects are manipulated in Smalltalk.

4.12 TECHNICAL DEFINITIONS

I have discussed three specific systems to provide an intuitive understanding as to what the five architectural levels in these figures mean. This intuitive understanding is entirely sufficient for nontechnical readers, who should immediately skip ahead to the next section. But for those who are technically inclined, it is important to realize that I'll not be using these five level names figuratively and loosely in this book. I will be using them as names for five definite and extremely specific innovations in software modularity technology with which every programmer is intimately familiar.

4.12.1 Process-level Objects

Definition: Process-level objects encapsulate state, behavior, and thread of control within an address space protected by physical address management hardware. Communication between objects is dynamically type-checked and dynamically bound, and entirely by exchange of values (exchanging references is prevented by physical address space barriers).

It is fortunate that most of this book will be focused at lower architectural levels, so that I'll rarely have occasion to use the term 'process-level object'. 'Appliance-level object' has exactly the right connotation by suggesting pre-assembled ready-to-use appliances like computers, printers, and disk drives, components that nonexperts can assemble by merely stringing some light-weight cable. However, I decided to avoid this term because it has entirely the wrong connotations in software. The world assumes that appliance-level object means personal computer software. So I eventually settled on 'process-level' object to mean the tightly encapsulated, loosely coupled objects of timesharing systems such as Unix.[7]

Systems that support process-level objects use physical means, such as address management hardware, to give each object its own address space, precisely as if it were running in a dedicated computer. Although this isolation prevents any possibility that one process might damage another accidentally, these address space barriers restrict beneficial communication as well. This also creates definite restrictions on what can be carried across process boundaries by interprocess communication channels such as pipes, signals, and the like. Since timesharing systems are used by many users simultaneously, they must guarantee that other users' processes cannot be affected if one user's process should go awry.

Thus process-level objects are encapsulated within unusually strong firewalls. Whereas objects that share a computer's address space can in principle share references freely (i.e., pointers, addresses of data structures in memory), process-level objects can share only values. This is why interprocess communication channels like pipes and files can carry values only between applications. Pointer-based structures such as linked lists, trees, or the chip-level objects of my previous book, cannot travel through pipes or be stored in files because the pointers are invalid except within the sender's own independent address space.

This level of modularity is most thoroughly developed today in timesharing systems such as Unix and VMS. Personal computer systems such the Macintosh do not support process-level objects, except in the degenerate sense that each PC is automatically its own independent address space. But if we broaden the meaning of 'personal computer' to include several of them interconnected by a network, process-level objects now become quite

[7] Bus-level object was also considered, but rejected on the grounds that buses are actually at the same level as cards. That is, if cards are the bricks, then buses are the mortar. Insofar as I've been able to discover, hardware engineering has no widely accepted term for the next-higher level of integration than cards.

feasible. Distributed computing, network operating systems, and client/server architectures are three names for 'supporting process-level objects on personal computers'.

Process-level objects are one of the hottest topics in the industry in this age of interconnected computers. Object technologies range from antiquated, but still useful, networking technologies such as FTP (File Transfer Protocol), to the worldwide web, to explicitly object-oriented technologies like CORBA (Object Management Group's Common Object Request Broker Architecture), OLE (Object Linking and Embedding, Microsoft), to OpenDoc (Apple), and so forth.

4.12.2 Task-level Objects

> *Definition: Task-level objects encapsulate state, behavior, and thread of control (stack area). But they share the same physical address space without the benefits (and liabilities) of physical address management hardware. Communication between objects is dynamically type-checked and dynamically bound, and may involve exchange of both values and/or references.*

Process- and task-level objects are coroutine-based objects. They each encapsulate an independent thread of control as well as the state and behavior of lower-level objects. But task-level objects share the same address space with each other, whereas process-level objects each reside in an independent address space. Whereas process-level objects are heavyweight processes, task-level objects are lightweight tasks.

Since they share the same address space, encapsulation is weaker. But on the other hand, binding can be tighter and communication much faster. Task-level objects can exchange complicated pointer-based data structures quite easily, as in the cut-and-paste paradigm of the Macintosh. But of course this all comes at a price. If one application goes wrong, it can (and generally will) bring the entire machine down with it.

Task-level objects are not based on the heavyweight processes of time-sharing systems, where each object receives an independent address space. They are based on the lightweight tasking paradigm of personal computer operating systems. Each task resides along with all others in the address space of the personal computer and can readily communicate with (and be damaged by) other tasks.

Whereas heavyweight communication channels such as pipes can carry only *values*, i.e., strings of ordinary bytes, the arrows in Fig. 4.5 can also carry *references*, i.e., pointer-based structures such as the chip-level objects featured in my previous book.

4.12.3 Chip-level Objects

> *Definition: Chip-level objects encapsulate their own state and behavior, but they share a common thread of control as a global resource. Communication between objects is dynamically type-checked and dynamically bound, and may involve exchange of both values and/or references.*

By long-standing convention, process- and task-level objects are the domain of operating systems. Now we're crossing a gradual slope into the domain of programming languages as this domain is conventionally called. That is, task- and process-level objects are coroutines and can communicate asynchronously with each other. From here on, all lower-level objects will communicate by invoking each other as subroutines[8] and communication is strictly synchronous.

Chip-level objects are a hybrid breed, like operating system (process- or task-level) objects in some ways but like programming language (block- or gate-level) objects in others. This distinction is best understood in nontechnical terms as the distinction physicists make between open and closed systems. In a closed system, nothing enters from the outside. The system is known and predictable as of the time it was closed. In an open system, the system must adjust dynamically to influences from outside.

The tangible world that we live in is an open system. New circumstances and objects turn up all the time, and they interact with each other in ways that no one could ever predict. Operating systems are open systems too. New process- and/or task-level objects are continually added and subtracted while the system is in operation. This is why the terms, dynamic binding and dynamic type-checking, appear in each of the definitions stated earlier. The objects in an open universe don't know in advance what other objects they might communicate with. So they must bind to their neighbors dynamically, whenever and wherever they encounter them. And they must check that the interaction is feasible dynamically. It is not possible to do this statically unless the system is closed.

4.12.4 Block- and Gate-level Objects

> *Definition: Block-level objects encapsulate their own state and behavior, but share a common thread of control. Communication can be statically type-checked by the compiler, and generally is by modern*

---

[8] Or equivalently, by having the compiler expand the subroutine's code in-line, as with a C macro or a C++ in-line procedure.

compilers. Binding may be delayed until as late as load time, but is complete by the time the program is in operation.

Definition: Gate-level objects encapsulate their own state and behavior, but share a common thread of control. Communication is statically type-checked by the compiler. Binding occurs entirely at compile time.

Anyone who has ever heard of object-oriented programming must also have heard of the controversies arising from different views of what the term, object-oriented, means. Unlike chip-level objects, which are carried to user-specified receivers through generic streams that don't know or care what they're carrying and will blithely transport whatever the user drops in, some languages take a more tightly coupled, less flexible, but safer and more efficient approach. Intra-object binding is not done primarily on the user's desktop as with the chip-level objects presented thus far. Instead, the type of every object must be known and declared at compile time, in order that compile-time type-checking be done in advance, not while the objects are in use on the user's desktop, but inside their developer's compiler.

Since this is analogous to the way silicon fabrication lines bind cell mask layouts from libraries of predesigned blocks to fabricate silicon chips, I'll call the kinds of objects whose connections to their neighbors can be specified in advance and type-checked by a compiler 'block-level' and 'gate-level' objects.

4.13 PROGRAMMING LANGUAGE COMPARISON

Since many programmers are uncomfortable with applying hardware engineering terminology to software, I do not expect that the specific terms I've used in this chapter will ever be accepted except within very narrow circles. We are thoroughly conditioned to focus on the differences between software and hardware. We forget that such terms are more about *people* than they are about hardware, and that *people* are something that both hardware and software indisputably have in common.

However, I confidently expect that terminology will someday evolve that will allow the diverse communicants who build and use software to discuss matters that transcend any particular architectural level without the rhetorical confusion that expends so much energy on language wars to this day. To assist in this evolution, Fig. 4.9 presents, without further comment, a way in which such a vocabulary could assist in making rational comparisons among three C-based programming languages that are often involved in particularly virulent language war disputes.

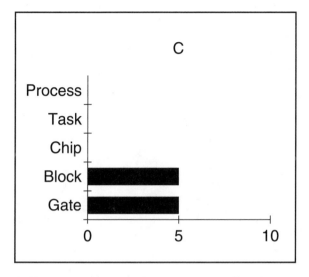

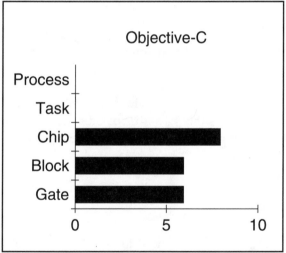

Figure 4.9 contrasts three C-based languages[9] according to how thoroughly they support object-oriented programming at the five architectural levels discussed in this chapter.

These particular figures consider only language functionality, not the functionality of associated libraries or operating systems. When these were included most or all of these languages would support higher-level kinds of

[9] For further details on Objective-C, see Cox and Novobilski, *Object-oriented Programming; An Evolutionary Approach*; Addison-Wesley; 1986. For information on C++, see Bjarne Stroustrup, *The C++ Programming Language*; Addison-Wesley; 1991.

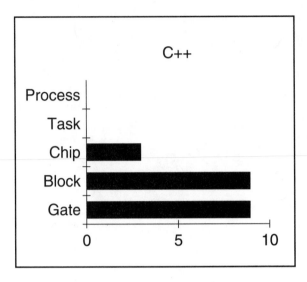

Figure 4.9   continued.

objects as well, depending on which libraries and operating systems you decide to include.

I encourage you to extend these figures to reflect the environment in your
shop. Such figures provide a rational alternative to the irrational, wasteful,
and tedious language wars that have dissipated so much energy during the
last decade, energy that might have been better spent preparing for the software industrial revolution.

4.14 SUMMARY

This chapter has argued that the compositional architecture of software is inherently heterogeneous because the people at various levels of a specialized
labor hierarchy have heterogeneous skills, interests, and needs. It has introduced a set of terms and definitions that people at different architectural levels can use to discuss objects at their level with people at neighboring levels,
without controversies that result from having only one word, 'object', to refer to the radically different components at diverse architectural levels.

Architectural levels correspond one-for-one with categories of people
who congregate at various levels of a specialized labor hierarchy, and these
come in an uncountable variety. The software community traditionally uses
a two-level hierarchy in which people come in only two varieties, programmers and users. We reflect this in the traditional distinctions between programming languages and operating systems.

Replacing today's two-level architecture (programming languages and operating systems) with a five-level one will be very helpful in this book. But counting or naming architectural levels is also a lot like counting or naming ants. Finding only one provides immediate grounds for concluding that there are more that you haven't seen and that looking harder will uncover more.

There is nothing magical about the five particular levels enumerated in this chapter. I have singled out these five only because they are already well known and supported by widely available programming languages, linkers, loaders, and operating systems, although not by the names I've assigned them in this chapter.

Chapter 5

Industrial Revolution

*Burns' Hog Weighing Method: (1) Get a perfectly symmetrical plank
and balance it across a sawhorse. (2) Put the hog on one end of the
plank. (3) Pile rocks on the other end until the plank is again perfectly
balanced. (4) Carefully guess the weight of the rocks.*

— Robert Burns

I have included this chapter because this book originated in the line of inquiry described here. To those who aren't immersed in the software engineering and computer science establishment, it will be a needless repetition of ordinary common sense reinforced with historical details about the industrial revolution. In this case, skip this chapter entirely. You have already absorbed its full meaning, which is that specification and testing tools are just common sense.

This chapter is for those who are fully immersed in the software engineering paradigm. As detailed in Chapter 3, the established paradigm is that there is no solution, even in principle, to the software crisis. Or in the words of Fred Brooks, there is no silver bullet capable of slaying the software werewolf, not even in principle. According to this established paradigm, electronic goods such as computer software are so utterly different from the tangible world of common-sense experience, that any comparison is at best an amusing analogy that can provide little insight into something as uniquely problematic as computer software.

Previous chapters have developed a very different paradigm based on a human-centric, as distinct from techno-centric, diagnosis of the causes of the software crisis. In this view, with the important proviso stated below, software is just another human activity. Thus comparisons with other human activities, such as manufacturing, are far more than mere analogies. They are the source of fundamental insight into how other humans have mastered equally complex tasks in other domains.

In this view, prior scientific revolutions such as the industrial revolution are not just historical anecdotes to read for amusement value before getting back to the "real" work of fabricating software from first principles as we've always done in the past. Prior revolutions are roadmaps, recipes for how other people have solved problems that seemed just as formidable in their day as software seems to us today.

5.1 SPECIFICATION, TESTING, AND LANGUAGE

Suppose that I insist that the period at the end of this sentence contains the software object of your dearest dreams. Suppose that I ask you to pay me for it? How would I prove that it does? How would you tell that it doesn't? How would I convince you to pay for my labor in producing it?

How do we expect vendors to ever build prefabricated components for others to assemble when the only way of answering these questions is by showing exactly how the components were fabricated (by distributing source code), or by distributing fully functional master copies (binary codes), and trusting that the customer will pay?

When I tell the clerk in a hardware store, "I'd like a pound of 1/2" carriage bolts," I can expect the clerk to know what I mean. I am conversing in a language that has evolved over millennia, a language for conversing about the domain that hardware stores and customers share.

This is one sign of the close relationship between specification and testing tools and the vocabulary of natural languages. The meaning of words in this vocabulary like "bolt" and "pound" are not determined through the formal "proof of correctness" so beloved by computer scientists; their meaning is established through informal specification and testing techniques.

As in the carriage bolt example, the experimental apparatus is usually no more sophisticated than our natural senses. If it looks like a 1/2" carriage bolt and the bag seems to weigh about a pound, I will probably assume that the bolts are acceptable, even to the point of using them for life-critical applications like hanging a swing on my porch. If my demands on the bolt are more specialized, such as joining booster rockets to a space shuttle, I might issue a more elaborate specification, and test compliance with more specialized equipment. The gauges I'll use can vary quite widely to ensure that the need has been met within whatever tolerance the customer is willing to accept and pay for.

But in software, this vocabulary has never evolved. We have never developed tools for detecting compliance between a specification and a putative implementation because the only thing that is concrete about software are the processes (programming languages) that we use to fabricate it from first principles. Names, such as "stack" or "document," refer to abstract data types, not concrete ones, since the only things that are concrete in software are processes, not products.

This absence of a product-centric focus in software is not inevitable. It is not some inescapable consequence of the essence of software as Fred Brooks argued in his "No Silver Bullet" paper. It is partially a sign of the youth and immaturity of software, and partially an outcome of the intangibility of a product that cannot be even partially specified or tested with the aid of our natural senses.

Most of all, it is the result of the same breakdown I've been pointing to throughout this book. Although we know that people are the sole source of electronic property such as software, we've never solved the problematics of buying, selling, and owning goods made of bits. Without a robust incentive for providing them, nobody has bothered to build whatever standardized components we might need. Since software is nonetheless in very high demand these days, everyone fabricates whatever is needed from first principles instead of by assembling standardized components from a market.

5.2 SPECIFICATION AND TESTING TOOLS

One of the paradoxes of software engineering is that, although concrete data types such as stacks are utterly invisible and undetectable by the natural senses, we've never developed instruments to make their static and dynamic properties visible. We have never developed instruments capable of measuring their static and dynamic interfaces analogous to scales, micrometers, and calipers that are routine in mature engineering domains. Without such instruments, there is no basis for a widely understood vocabulary, comparable to my hardware store analogy. So long as only the processes (computer languages) for building software are standardized, and not the products of those processes, we've no choice but to fabricate software from first principles. For without such instruments for detecting compliance to a standard, how can we possibly construct an industry based on assembly of interchangeable parts?

Assembling software from prefabricated components implies that these components exist, not merely as abstract data types but as concrete ones, tangible entities that someone has implemented and tested and made available in return for some valuable incentive. At the very least, we'll need names for these components, and tools to compensate for the inability of our natural senses even to detect them. The nature of these specification and testing tools is the topic that will concern us in this chapter.

Tools for specifying the external interfaces of software components and for the gauging compliance of implementations to specifications do not exist today except in the most rudimentary form. Such tools are certainly not integrated with the software development toolkit in the way that text editors, compilers, and linkers are.

Specification and testing tools are not common for the same reason that we don't use a ruler to cut string to tie a package. Cut-to-fit activities need only raw materials (string) and an implementation tool (scissors). Such tools become relevant when the transition to an industrial approach based on standardized, interchangeable components has been made. If the job is manufacturing 24" shoelaces, I'll need a specification tool to quantify the length to be implemented. If my implementation process is imperfect, I may also need a testing tool (perhaps the same ruler) to verify that the length implemented was as specified within some acceptable tolerance for deviation.

Nonetheless, the topic is still one of the key steps of moving to a true engineering approach that emphasizes assembly of standard, interchangeable off-the-shelf components. So long as our only tools are implementation tools, we'll have abstract data types but no concrete ones. We can define standard processes such as programming languages and methodologies, just as cut-to-fit craftsmen define standard processes such as filing, sawing, and drilling. But the standard product catalogs will be absent, and thus the center around which true engineering revolves. The consumer's external viewpoint, as reflected in a specification, will remain irrelevant so long as the software universe revolves around a solitary handcraftsman wielding cut-to-fit implementation tools such as compilers.

Precisely because of this feeling of irrelevance, it is crucial to know that others have overcome similar obstacles in the past. The example that I will present in this chapter is the period in the industrial revolution during which the U.S. armament industry adopted interchangeable components. However, the similarities between their case and ours does not mean that the obstacles are identical. Although the pace of change is faster today, the obstacles to achieving an analogous revolution for software are serious indeed.

The industrial revolution involved the production of tangible goods, physical objects that could be sensed with natural senses with which both producers and consumers were equally equipped. Both the process and the resulting product were painfully apparent to any observer. Neither is true of intangible goods such as electronic data and software. Even so, the gunsmiths found that the precision that interchangeability required exceeded that available from the unaided senses. Their pioneering attempts to build interchangeable components with implementation technologies alone failed. It was nearly a quarter century before they realized that they also needed to make equal or even greater investments in specification/testing tools.

The most fundamental point of divergence is that industrial revolution goods were tangible assets, things that can be robustly bought and sold by the copy. This is certainly not the case with electronic goods that can be replicated and transported at the speed of light through information-age infrastructures such as networks. This is the most fundamental obstacle of all because, so long as it is not surmounted, it undercuts the revenue-generating engine, the market forces that drove the manufacturing age to such monumental successes (and, of course, equally monumental abuses).

5.3 ARMORY PRACTICE

> *Much of the excitement generated by the special investigations of 1826 can be traced directly to Hall's success in combining men, machines, and precision-measurement methods into a practical system of production. In this sense, Hall's work represented an important extension of the industrial revolution in America, a mechanical synthesis so different in degree as to constitute a difference in kind.*
> — Merrit Roe Smith, *Harpers Ferry Armory and the New Technology*

Revolutions are times when paradigms change, when an old exemplar is discredited and a new one takes its place. In order to anticipate how the software industrial revolution might unfold, we now turn to the industrial revolution. In particular, we turn to that part of the story during which mankind discovered that high-precision interchangeable parts were actually, all things considered, *less* expensive than the low-precision cut-to-fit components that had dominated the craft-centric approaches of the past.

This part of the industrial revolution occurred primarily in the New England region of the United States about two centuries ago. These innovations played a considerable role in transforming this country from a colonial backwater of Britain to a position of economic influence and power that has only begun to be seriously challenged within the last quarter century or so. The history of this transition is uniquely accessible. The Springfield Armory of Springfield, Massachusetts, the Ely Whitney Museum of Hamden, Connecticut, the American Precision Machinery Museum of Windsor, Vermont, and the Smithsonian American History Museum in Washington, D.C., all provide vivid examples of the transition from a craft-centered approach, centered around fabrication of cut-to-fit components, to the industrial approach that is so much a part of modern life.

For a retrospective of how things were prior to this period, I recommend the Colonial Williamsburg Museum of Williamsburg, Virginia, for a fascinating reconstruction of life in colonial times. Costumed employees demonstrate life in a medium-sized town, long before the industrial revolution changed this way of life forevermore. The gunsmith shop in the town demonstrates how muskets were built before the industrial revolution, just as we build software today. When I was last there, the gunsmith was making a wood screw to join the wooden stock to a barrel. He began the day at the forge behind the shop, forging wrought iron to make an iron rod. He spent the rest of the morning at his vise, filing this rod into the head, shank, and threads of a wood screw, working entirely by hand and by eye.

It is impossible not to admire this gunsmith, not only for his skill and the quality of his work, but also for the conviction of his beliefs. He takes an uncompromising stand in favor of fine handcraftsmanship. Faced with the trade-offs of his imperfect world, undaunted by the soot, imperfection, rust, and rot, he is acutely aware of the trade-offs between cost versus performance, quality versus quantity, build versus buy, flexibility versus strength. Who cannot admire this uncompromising bastion of a value system that has since been relegated to hobbyists and museums?

There is also a darker side to this picture. His craftsmanship was so exquisite that I asked if I could buy the musket. "I generally deal with museums, not individuals, but I could consider it if you're interested," he said. "I'm asking $30,000. Check back in a year or two." Every trade-off he made favored his own system of values, not those of potential customers such as myself. Who can also not appreciate that definition of 'quality' might not coincide with that of their consumers? Issues that such customers might care deeply about, cost and schedule for example, didn't even impinge on his priorities. A $30,000 muzzle-loading musket might be high quality to an antique arms collector, but it would be a very low quality for a soldier who might care far more about firing rate or accuracy. It would matter even less to a quartermaster concerned with whether he will be able to replace worn or broken parts in the field.

The cottage-industry approach to gunsmithing was in harmony with the economic, technological, and cultural realities of colonial America. It made sense to expend cheap labor as long as steel was imported at great cost from Europe. But as industrialization drove materials costs down and demand exceeded what the gunsmiths could produce, they began to experience pressure to replace the cottage-industry gunsmith's process-centered approach with a product-centered approach: high-precision interchangeable parts to address the consumer's demand for less costly, easily repairable products.

The same inexorable pressure is happening to the software industry as the cost of hardware plummets and demand for software exceeds our ability to produce it. As irresistible force meets immovable object, we experience the pressure as the software crisis: the awareness that software is too costly and of insufficient quality, and its development nearly impossible to manage. Insofar as this pressure is truly inexorable, nothing we think or do can stand in its path. The software industrial revolution will occur, sometime, somewhere, whether our value system is for it or against it, because it is our consumers' values that govern the outcome. It is only a question of how quickly, and of whether we or our competitors will service the inexorable pressure for change.

Although the old-time gunsmiths were the conservatives of the industrial revolution, it is not really accurate to say that they 'resisted' and 'lost' to the progressives. The gunsmiths barely resisted at all. They were secure in their profession. They knew that the progressive view was nonsense. They knew that machines operated by unskilled labor could not possibly compete with their finely honed craftsmanship. They knew that it would be impossibly expensive to obtain the close tolerances needed to interchange parts across firearms. They were too busy to even explain anything so obvious to the bureaucrats at the Bureau of Ordnance, let alone resist. The gunsmiths stayed busy in their workshops, filing away on their iron bars, while the world marched past and left them there.

5.4 THOMAS JEFFERSON

Judging from the following letter from Thomas Jefferson to a friend in 1785, it was actually the president-to-be of the nation who discovered the solution in the workshop of one of Napoleon's inventors, a gunsmith named Honoré Blanc:

> *An improvement is made here in the construction of the musket which it may be interesting to Congress to know, should they at any time propose to procure any. It consists in making every part of them so exactly alike that what belongs to any one may be used for every other musket in the magazine. The government here has examined and approved the method and is establishing a large manufactory for this purpose. As yet the inventor had completed the lock (the assembled parts of the trigger mechanism) only of the musket on this plan. He will proceed immediately to have the barrel, stock and their parts executed in the same way. Supposing it might be useful to the U.S. I went to the workman. He presented me with the parts of 50 locks taken to pieces and arranged in compartments. I put several together myself, taking pieces at hazard as they came to hand, and they fitted in the most perfect manner. The advantages of this, when arms need repair, are evident.*
>
> *— Thomas Jefferson*

5.5 ELI WHITNEY

In 1798, six years after he invented the cotton gin and thirteen years after Jefferson's letter, the federal government signed a momentous contract with Eli Whitney. Whitney agreed to deliver four thousand assembled muskets within a year and a half. This was itself unusual because the government's armories had never turned out anywhere near this number. Even more remarkably, Congress agreed to subsidize the work with payments in advance, at a time when private contractors were customarily paid only on delivery. And strangest of all, the largess flowed to a man who was close to bankruptcy from trying to defend his patents for the cotton gin, and who knew nothing of making guns!

Whitney was, to put it bluntly, well-connected. Or in a more positive light, he knew how to listen to what customers wanted and to promise what they were willing to buy. However, by the end of 1800 Whitney had delivered few of the four thousand muskets and was in danger of running out of cash.

The idea of building machinery capable of producing interchangeable parts seems to have come to him about this time. The idea did not originate with him. It was probably suggested to him by Thomas Jefferson, who probably acquired it from Honoré Blanc. He no doubt got it from someone else. Although no one had yet made guns from interchangeable parts, lockmaking was a well-established industry that involved considerable specialization of labor and interchangeability even in the early 1700s.

However, in 1801, when it was clear that he'd be late and would need more money, Whitney duplicated the show that Blanc had given to Jefferson, this time to President John Adams, President-elect Thomas Jefferson, and other leading politicians. From that day forward Whitney was a hero, and so he remained until modern scholars at the Smithsonian Museum discovered the original components of his demonstration in the archives. Examined under modern microscopes, it became clear that Whitney had staged his demonstration. He had provided hand-fitted specimens prepared specifically for this occasion! It was not until 1808, 7.5 years late on his 1.5 year contract, that he delivered the last of the 4000 muskets! Even then, Whitney never delivered on his promise to provide mass-produced guns with interchangeable parts!

In other words, Eli Whitney was a better salesman than he was a technician! But does this mean he was a charlatan? His contemporaries certainly didn't think so, nor did the War Department, who continued to call on him for advice. Roswell Lee, an early assistant who later became superintendent of the Springfield Armory and later brought Whitney's vision close to fulfillment, remained a close friend and admirer. The historian, Harold Livesay, wrote this judgment of Whitney's contribution:

*Among these pathfinders, Whitney ranks first, but not because he
originated interchangeability, perfected it, or applied it successfully to
mass production; for he did none of these. What he did do was to
perceive identical parts as a precondition to volume production. He
applied his energy and genius to the problem, and he influenced his
contemporaries and successors who addressed the same challenge.
Whitney built the first arch in the engineering bridge that spanned the
gulf between the ancient world's handicraft methods of production and
the modern world's mass manufacturing, a bridge across which his
countrymen and the rest of the developed world marched to prosperity.*

So, in spite of his shortcomings as a technician, Whitney made the one
contribution that was truly indispensable. He identified clients whose problems could be solved in no other way, convinced them that the obstacles
could be overcome, and sustained their confidence through the 24 years of
failure it took to work out the details. Whitney won his place in history, not
because of his technical achievements, which actually were substantial but
never came together in the right combination. Whitney's contribution was to
create the climate in which his successors, notably Roswell Lee at the armory
in Springfield and John Hall at the one in Harpers Ferry, could succeed.

Whitney was a cotton gin inventor. Roswell Lee was a soldier, a Lieuten-
ant Colonel in the War of 1812 prior to being appointed superintendent of the
Springfield Armory. John Hall was an entrepreneur and inventor, who ran a
cabinet-making shop before being assigned space in the Harpers Ferry armory to develop a technique for manufacturing his patented breech-loading
rifle. Neither was trained, which is to say indoctrinated, in the craft of gun-
making, except for Hall, whose motivation came from his desire to sell his
rifle to the government. Arrayed staunchly against them were the cottage-
industry gunsmiths described in this quotation, championed by James
Stubblefield, the superintendent of Harpers Ferry armory and Hall's immediate supervisor:

*Over the years such thoroughly inbred and highly individual work
habits served to hinder rather than encourage innovation at Harpers
Ferry. Because so many armorers and supervisors had been reared
according to the conventions of the craft ethos, they found it extremely
difficult to adjust to the increasingly specialized demands of industrial
civilization.*

*Except for using commonly known forging, grinding, polishing,
boring, and rifling machines, they relied mainly on the dexterous use*

of hand tools to perform their work. Unlike their more flexible contemporaries in New England, they possessed a high degree of manual skill and saw no need to compensate for what little they lacked by cultivating mechanical know-how.

Above all, they considered themselves artisans not machine tenders, and as such, believed in the dictum that an armorer's task consisted in making a complete product: lock, stock and barrel. Ideologically speaking, then, they had neither the preparation nor, it seems, the inclination to introduce new techniques at the armory. In a sense, they were too skilled, too artistically inclined, too satisfied with the way things were, to sanction change or contemplate seriously its possible benefits.

Merrit Roe Smith, *Harpers Ferry and the New Technology*

The key ingredients of John Hall's success stand on display in the National Museum of American History at the Smithsonian Institution in Washington, D.C. As I will frequently use these tools to typify what the software industry has yet to accomplish, it will be useful to consider two representative examples in some detail. But before turning to the tools, Fig. 5.1 shows the object to which these tools were oriented, John Hall's breech loading rifle of 1819.

Notice the complexity of this weapon compared to the simplicity of its muzzle-loading competition, and even the cartridge-based breechloaders of today. John Hall's incentive for working so hard on machine-based manufacturing technologies was that without them, his design didn't stand a chance. In fact, this turned out to be fatal in spite of the fact that he succeeded in making them. Infantry commanders could never be persuaded that the delicate-appearing mechanisms would hold up under hard use in the field.

It took Hall four years to fill the army order because he had to start nearly from scratch. He brought in mechanicians, including Thomas Blanchard, whom Roswell Lee had sponsored at Springfield to build the pattern lathe shown in Fig. 5.2, to build machines to shape the various parts of the rifle. This water-powered machine could follow an irregularly shaped pattern to mill a rectangular piece of wood into an all-but-finished gun stock.[1]

This primitive tool was the analog predecessor of the numerical milling machine, the high-tech digital successor that plays such a vital role in modern manufacturing. This tool typifies the state of software engineering today.

---

[1] For anyone wondering where all these analogies are leading, this machine automatically compiled a high-level specification of the problem into an all-but-finished implementation of that specification. Is everyone with me now?

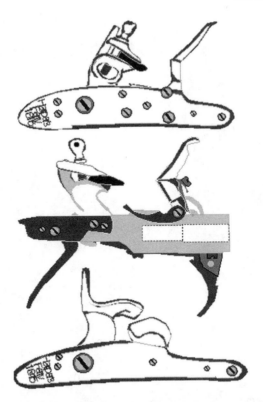

Figure 5.1 This figure shows the lock and breech assembly of John Hall's innovative breechloading flintlock musket of 1811 (middle). Contrast the complexity of Hall's lock (middle) with previous (top) and succeeding (bottom) arms. To overcome this deficit, Holt successfully sold the army on, not just the merits of breechloading arms as a weapon, but on the dream that 'if a thousand guns were taken apart and limbs thrown promiscuously together in one heap, they may be taken promiscuously from the heap and will all come out right'. In March 1819, with the support of John C. Calhoun, Hall signed a contract to produce 1000 of these rifles at the Harpers Ferry armory, the entire project to be subsidized by the army, including Hall's salary of sixty dollars a month.

Before such tools were invented, stockmaking was a highly skilled craft. Tools of risk such as rasps, saws, spokeshaves, and chisels were used to incrementally shave away the waste until nothing but a finished gunstock was left. Pattern lathes, by contrast, are tools of certainty. They allowed even unskilled immigrant laborers to turn out finished gunstocks of equal quality as fast as they could feed in more wood.

Figure 5.2 In the Blanchard pattern lathe, a pattern stock at the back of the machine rotates at the same speed as the workpiece at the front. A wheel traces the pattern to govern the separation between a cutting wheel and the workplace. The tracing wheel automatically moves along the length of the pattern, causing the cutting wheel to reproduce the pattern on the workpiece. The machine automatically disengages its drive belt when the carriage meets a stop at the end. (Courtesy of National Museum of American History; Division of History of Technology, Smithsonian Institution)

Although the replication rate of the Blanchard lathe was hardly comparable to the rate at which software copies can be replicated through a network, the impact on the gunstock makers was so devastating that this craft eventually disappeared almost altogether. The main difference in software is that software subcomponent providers have been in the same devastating situation from the very beginning, so no analogous community has even formed.

5.6 ROSWELL LEE

In spite of the obvious impact of implementation tools such as this pattern lathe, the Smithsonian exhibit was organized to emphasize that such tools were only necessary, not sufficient, for in the same exhibit stands the box of inspection gauges shown in Fig. 5.3. Although the displacement of tools of risk by tools of certainty was a part of the interchangeable parts story, it was not the dominant part. The stage of evolution represented by these inspection gauges has not yet occurred in software.

One of the crucial discoveries of the initial quarter-century of experimentation, frustration, and failure was that, as the primitive steels of that era wore down and the wooden machine frames crept out of alignment, the parts they produced steadily fell out of tolerance. Although making parts by machine sped production and reduced the need for skilled craftsmen, the Ordnance Department insisted on strict interchangeability regardless. At its insistence Roswell Lee inaugurated a system of gauges that by 1819 had reached a point of considerable sophistication. A visitor to Springfield that year describes how the system worked:

> *The master armorer has a set of patterns and gauges. The foremen of shops and branches and inspectors have each a set for the parts formed in their respective shops; and each workman has those that are required for the particular part at which he is at work. These are all made to correspond with the original set, and are tried by them occasionally in order to discover any variations that may have taken place in using them. They are made of hardened steel. If this method is continued, and the closest attention is paid to it by the master workmen, inspectors, workmen, and superintendent, the designed object will finally be obtained.*

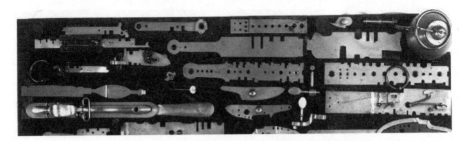

Figure 5.3 Although powered implementation goals like Blanchard's lathe were clearly an important innovation, it was the introduction of inspection gauges such as these that made truly interchangeable parts possible. The ones shown here are from the Springfield Armory. (Courtesy of National Museum of American History, Division of History of Technology; Smithsonian Institution)

It is significant that the straightforward meaning that these inspection gauges give the term, 'specification', is rarely used in today's software circles. It is specification/testing tools like these mundane inspection gauges that are the strategically vital, but as yet unavailable, technologies that must yet be invented before the software revolution can be anything but an impossible dream.

5.7 REVOLUTIONS DON'T HAPPEN OVERNIGHT

To highlight the importance of this point, and to comprehend the magnitude of how much work remains before these lessons can be applied to software, the time line in Fig. 5.4 shows the progression of the key events of this period.

This revolution hardly happened overnight. In 1799 there began a period of invention in which Congress poured money into a bottomless hole, getting nothing in return except more promises. In 1822 John Hall at Harpers

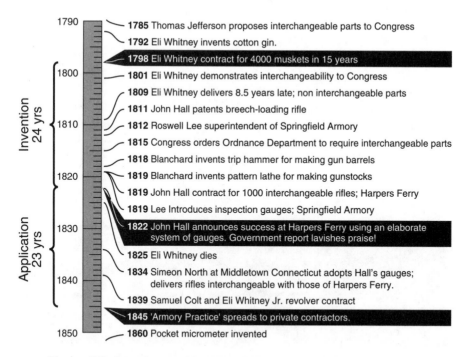

Figure 5.4 Interchangeable parts did not come about easily or quickly. Changes of this magnitude seem to follow a characteristic time constant that is rarely less than a generation or so. The steamboat, railroad, automobile, farm machinery, recording, telephony, and electric power industries involve similar periods of time. Will software evolve any faster?

Ferry, with copious assistance and inspiration from Roswell Lee at Springfield, managed to overcome these challenges in a pilot-scale project. Even then, an additional quarter-century was to elapse before the techniques spread beyond the government armories, first to private contractors and, ultimately, to industry as a whole.

In other words, revolutions do not occur overnight. This one involved a 24-year-long period of invention, a bleak and profitless period dominated by failure until John Hall scored the first clear pilot-scale success at Harpers Ferry. Then there was another 23-year-long period during which the idea slowly spread to other sites. Finally there was the indefinitely long period during which the idea established itself as the dogma of the new era. Interestingly enough, it was only during this expansionary period that truly adequate tools were developed, notably the pocket micrometer.

Another key point is that revolutions do not obsolete all that came before them. Blanchard's lathe did not obsolete the spokeshaves, saws, and rasps that had heretofore been used to build gun stocks. I can still find hand tools in my local hardware store that are no different from the ones colonial craftsmen used. On the other hand, early versions of tools of certainty such as Blanchard's lathe, can be found only in museums. I shall not be surprised to find that simple hand tools like Cobol, Fortran, and C will be with us forever, while the earliest incarnations of tools that will displace them evolve so quickly that they rapidly disappear. Tools evolve so quickly during revolutions that the original incarnations are soon forgotten.

5.8 THE HEROES OF THE INDUSTRIAL REVOLUTION

Finally, and most significantly, notice who the true revolutionaries were, the people that really made this revolution happen.

> *"It does not diminish the work of Whitney, Lee, and Hall to note the relentless support that came from the government, notably from Colonel Wadsworth and Colonel Bomford in the Ordnance Department and from John C. Calhoun in Congress. The development of the American system of interchangeable parts manufacture must be understood above all as the result of a decision by the United States War Department to have this kind of small arms whatever the cost."*

Where were the reactionaries while all this was going on? Smugly certain that machines could never build something as complicated as a gun, the cottage-industry gunsmiths stayed busy in their workshops, filing steel bars to make hand-fitted parts. The world marched past and left them there!

But the industrial revolution is nonetheless relevant for what it teaches us about what Merrit Roe Smith attributed John Hall's success at Harpers Ferry armory to: *"...combining men, machines, and precision-measurement methods into a practical system of production; a synthesis so different in degree as to constitute a difference in kind."*

Men[2]: Revolutions are about *people* and their paradigms even more than they are about technologies. This story is about how people respond to the crisis that triggers the emergence of new paradigms. As we shall see, the role of the establishment in this tale was played by the cottage industry gunsmiths. They were hardly the heroes of this revolution. The heroes were a completely different group, the gunsmiths' customers. In frustration with the intransigence of the establishment, they established a new set of conditions that encouraged a new group to step forth, early industrialists such as Ely Whitney who were prepared to put their customers' idea of what was important ahead of their own preconceptions.

Machines: This tale demonstrates the evolution of implementation technologies away from tools of *risk* of the cottage industry craftsman to tools of *certainty* such as the numerically controlled machine tools of today. However, I shall not emphasize this aspect as extensively as the other two, because the value of process-centric, i.e., language and methodology, improvements are already well-accepted in software. The progression from assembly language, to second generation languages like Cobol, to third generation fabrication languages like C, to object-oriented assembly technologies like Smalltalk, to logic/mathematics intensive languages[3] like Z and VDM, is a well-established theme in our world.

Precision measurement methods: The tale demonstrates a growing appreciation of the value of *standards* and of product-centric specification/testing technologies for enforcing them. I am referring to the engineer's equivalent of the scientist's telescope, microscope, and butterfly net: the micrometers, rulers, and inspection gauges that have become the critical path in manufacturing now that numerically controlled machine tools have made steel almost as easy to replicate as software.

The paradigm shift of this period had fundamental repercussions throughout every level of society today. Some of them are relevant to software, but many of them are not. Software development is not, and will never be, an assembly line operation as in gun or automobile manufacturing.

---

[2] The histories I've studied do not show that women played a dominant role during this period, other than presumably in the home and in providing much of the labor force in the woolen and cotton mills.

[3] Unfortunately, much of the work on using logic/mathematics-based languages is being developed under the name 'specification technology' today. I'll be using the term 'specification/testing' technology in quite a different sense throughout this book, to mean a technology that plays no role whatsoever in implementing an object, but only in specifying what the object should be. A lathe is an implementation technology for building round things. A caliper is a specification technology for specifying round things.

5.9 SOFTWARE ENGINEERING

*CASE stands for Computer Aided Software Engineering; it can be
used to mean any computer-based tool for software planning,
development, and evolution. Various people regularly call the following
'CASE': Structured Analysis (SA), Structured Design (SD), Editors,
Compilers, Debuggers, Edit-Compile-Debug environments, Code
Generators, Documentation Generators, Configuration Management,
Release Management, Project Management, Scheduling, Tracking,
Requirements Tracing, Change Management (CM), Defect Tracking,
Structured Discourse, Documentation editing, Collaboration tools,
Access Control, Integrated Project Support Environments (IPSEs),
Intertool message systems, Reverse Engineering, Metric Analyzers.*
— What is CASE; Answers to frequently asked software engineering questions;
Internet news group, comp.software-eng

The relevance of the industrial revolution story to software lies in how it
influenced the build-to-order, assembly-intensive industries that it enabled,
not how it influenced the fabrication-intensive industries where it was pioneered. Software is reasonably certain to remain a build-to-order business,
unlike the assembly lines of gun or automobile manufacturing.

Someday, possibly as much as a generation from now, when software is as
mature as manufacturing is today, process-centric issues such as which language a component was written in, will recede into the background and be
replaced by product-centric issues such as to which specification a component complies. Languages will not dominate interface considerations since
the dominant tool will be the product, not the process.

This goal is unrealistically ambitious today. Everything we do in software
revolves around programming languages and not around standards for interchangeable components, so every programming language imposes its
own unique conventions. This section will concentrate on a far less ambitious, and far more relevant, example of where lack of interchangeability and
standardization is already a problem today. This is interchangeability of
components across different releases of a library within a specific object-
oriented programming language. We will leave interchangeability across
languages aside for the future and concentrate on interchangeability as components evolve over time.

When we first started building Objective-C class libraries at Stepstone a
decade ago, our productivity was initially high. These first few months were
our cut-to-fit craftsmanship phase. Productivity was high because every
new line of code needed to be compatible only with the code in its immediate environment. But after we'd shipped our libraries to customers, external
dependencies began to impinge. These dependencies resulted from the way

our customers' code used our classes, something that we obviously had absolutely no control over. The growing numbers of external applications began to impose a de facto specification that we needed to comply with every time we touched our code in any fashion, such as during every repair, extension, or port.

Although it took us some time to realize what was happening and do something about it, the symptoms were depressingly clear. We noticed that our productivity was no longer so high. The most costly symptom was in porting graphical user interface components from machine to machine. Porting is clearly something that should affect only an implementation detail (the hardware platform), but leave the abstract specification completely unchanged. But since porting graphical user interface components from, say X-Windows on one platform to bare graphics hardware on another, can be highly complex, there was considerable room for uncertainty as to whether our software components presented the same interface to their client applications (not to mention to application users) as we ported it from platform to platform.

The underlying problem was that our customers expected standardized components; i.e., components that could be interchanged from release to release. But the standard to be complied with was embodied solely in our customers' applications. Not only was this standard not under our control, we no longer even knew what the standard was. No matter how trivial the change, or how obvious the bug, there was always a possibility that some customer's code relied on things working precisely as in the older releases. The result was a depressing fall in our own productivity and, of course, the demoralization that comes from being aware of numerous bugs and opportunities for improvement but being caught up in a closed system with no room for headway.

5.10   SPECIFICATION AND TESTING TOOLS

What is actually needed is to capture the specification explicitly in a specification tool in some fashion that is independent of any particular implementation. This is the role of specification and testing tools, software analogs of the measuring devices and inspection gauges of modern manufacturing. Oddly, the academic establishment assumes that this is impossibly hard, that specification and testing tools are some kind of holy grail for researchers to aspire to but not to attain. We forget so easily that craftsmen less sophisticated than programmers do precisely this all the time, every time they judge a board for squareness with the aid of a carpenter's square instead by comparing it with another sawed board (Fig. 5.5). The carpenter's square, the

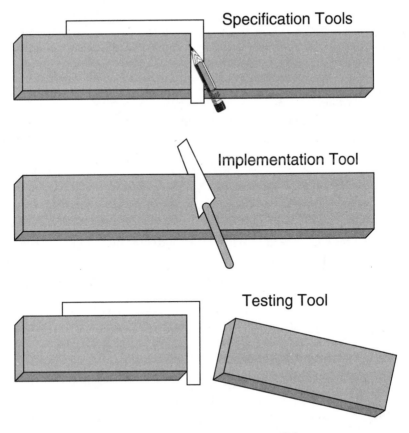

Specification Tools

Implementation Tool

Testing Tool

Figure 5.5 shows the relationship among specifying, implementing, and testing tools as I'll use those terms in this chapter. In the case shown here, the same carpenter's square is used for both specifying and testing, a sign of the intimate relationship between these two activities.

electrician's measuring tape, machinist's micrometer, and the butcher's meat scale are all exemplars of exactly the kind of tool we're seeking, tools for specifying the externally visible properties of some object without regard for any particular implementation.

I imagine the reason we assume this is hard is that we've adopted an impossibly high goal. The very word "specification" is automatically burdened with one of two meanings in software both of which are extremely, perhaps even impossibly, hard to ever achieve. One of these meanings is that the only proper goal of a specification tool is serve as a software tool of certainty, capable of automatically transforming an abstract specification into an implementation without any possibility of human error. The second, which is obviously closely related, is "formal proof of correctness"; i.e., demonstrating that an implementation, however produced, complies with its specification. Both uses of this term obscure the absence of true specification technologies in software—tools like the rulers, protractors, calipers, micrometers, and gauges of manufacturing. Programmers continue to rely exclusively on implementation technologies, be they ultrahigh-level programming languages—tools of certainty like the Blanchard lathe—or lower-level languages—tools of risk like the rasps, files, and spokeshaves that it displaced.

Software engineering has not yet reached even this primitive level of refinement. The search for better software productivity and quality is dominated by the movement to ever more powerful implementation technologies. We still measure advancement in this industry by the movement from assembly language to structured languages to object-oriented languages and so forth. But this advancement amounts to improved tools for fabricating cut-to-fit components from first principles, initially by hand and later by machine.

Specification tools, as I'll use this term here, have no implementation capability at all. The carpenter's square is solely devoted to specifying what is to be implemented, leaving that task to implementation tools such as a saw.

I am not adopting this unambiguous and low-tech definition to diminish the importance of tools of certainty in relation to the tools of risk in this figure. Fifth generation programming languages based on formal mathematics, for example, may someday prove to be as helpful in software as miter boxes and radial arm saws are in carpentry, since these are all ways of automatically transforming a specification into an implementation with greatly reduced opportunity for error. In the long run this low-tech meaning is potentially even more important in view of what this new meaning implies with respect to changing the relative unimportance of specification tools with respect to implementation tools today.

5.11   WHAT IS A STACK?

Consider the most hackneyed object-oriented example of all, the ever-popular stack class. Everyone understands stacks so long as we keep the discussion on the plane of abstract data types, not concrete ones. If we think of stack as a *concept*, an abstract data type that every programmer must fabricate in a cut-to-fit manner each time one is needed, issues like standardization and interchangeability don't arise. A stack is an abstraction with two abstract capabilities named push and pop. When we've pushed elements onto a stack, pop returns them in last-in, first-out order. What could be simpler than that?

When we turn to considering stack as a *concrete* data type, a prefabricated *thing* and not a *concept* for everyone to fabricate as needed, we must be specific about details that the abstract data type approach resolves entirely through cut-to-fit adjustment. The simplest issues involve static matters such as precisely how the method names are spelled. Are they spelled push and pop as in C++? Or are they spelled push: (with a colon) and pop as in Objective-C? Or is it add and remove as in Bertrand Meyer's rationalized naming scheme for the container classes in Eiffel?[4]

Compile-time typechecking is solely concerned with static naming issues such as these, but the issues get harder as soon as we recognize that dynamic issues are just as important to a consumer. They are not only worried about whether the new release will compile correctly, but in whether the new stack release will *run* correctly when they link it with pre-existing applications. But since compile-time checking addresses only static issues of method naming, what tools do we use for dynamic ones? What tools do we use to verify that push: and pop exhibit the same dynamic properties that they had in the stack class that was shipped in the previous release?

Did you notice how quickly this discussion slipped from the specification question of *what* to questions of *how;* to considering which programming language was used to fabricate and invoke the stacks of this example? The software universe revolves around the cut-to-fit craftsman's compiler and everything else revolves around that. There is no such thing as a stack in any concrete sense. In the present process-centric state of software, everything revolves around which specific process was used in building a component and not around product-centric issues, such as the specification of the component itself.

This process-centric orientation is characteristic of immature industries like software (and preindustrial crafts in general). The priorities are completely reversed in mature industries such as carpentry. When we specify a stack, our thoughts immediately gravitate to process-centric questions of whether the stack was written in C++, Smalltalk, Objective-C, or Eiffel. But when a carpenter specifies a 1/2" carriage bolt, which process was used in making the bolt is entirely immaterial. The producer is free to use any process for building bolts so long as it delivers acceptable 1/2" carriage bolts. Even hand-filing them from steel bars is acceptable if you think you can make money that way.

---

[4] Bertrand Meyer, *Tools for the New Culture: Lessons from the Design of the Eiffel Libraries;* September 1990 CACM.; Vol. 33, no. 9, pp. 40-60.

5.12 ASSERTION CHECKING

I want to switch now from top-down arguments as to why specification and testing technologies are needed to a bottom-up description of how they could be provided. The primary base capability we'll need to build inspection gauges that stand apart from, but that sample the static and dynamic properties of an external concrete data type such as my stack example, is readily available within every programming language. This base capability is simply (1) the ability to invoke an external component in a controlled, repeatable fashion, and (2) observe the results to ensure that they are as expected.

These capabilities exist within every programming language. Eiffel provides assertion checking in an unusually well-thought-out fashion that is closely integrated with the rest of the language. Fortran, by contrast, provides only generic capabilities of subroutine invocation and ordinary conditionals, but these turn out to be precisely the base capabilities that are needed to build perfectly acceptable assertion-checking capabilities. The same is true of Smalltalk and C, but in the case of C, some long-forgotten programmer already went to this trouble. His solution exists in every C environment I've ever used as a readily available macro definition file, /usr/include/assert.h.

This macro, assert(), takes an expression as its only argument. It merely verifies that the expression is true, in which case execution continues normally, just as if the assertion were not present. If the expression is false, the macro prints the file name and line number of the line containing the invalid assertion and triggers a fatal exception. For example, suppose line 12 of file Foo.m contains the assertion

```
assert(2+2 == 4);
```

Execution will continue normally because this assertion is true. But had the line read instead

```
assert(2+2 == 5);
```

the following error message would be printed and execution would be interrupted.

```
assertion error in file Foo.m line 12
```

This simple capability turns out to provide exactly the foundation that is needed for a number of specification/testing approaches. These will be considered in the following sections. But first let me show how assert() is actually implemented in C and the convenient (but nonessential) consequences of defining it as a macro instead of a subroutine.

```
#ifndef NDEBUG
        extern void assertionTrap(char* filename, int
lineNumber);
#       define assert(ex) { if (!(ex))
assertionTrap(__FILE__, __LINE__); }
#else
#       define assert(ex) /* empty */
#endif
```

By default, NDEBUG is not defined, so the first definition applies. Since 2+2 == 5 is false, (!(2+2 == 5)) is true. Thus assert(2+2 == 5) would invoke the assertionTrap() subroutine. This subroutine merely prints an error message and stops execution by raising a FalseAssertion exception.[5] The two arguments of this subroutine, __FILE__ and __LINE__, pass the file name and line number of the failed assertion to this subroutine so that they can be printed in the error message to help in fixing the problem.

The only real significance of defining assert() as a macro instead of as, say, an ordinary subroutine, is that the overhead of evaluating the if() statement at runtime can be eliminated by setting the preprocessor symbol, NDEBUG and recompiling the program so that the second definition of assert() applies. In this case, the assert() statements compile into nothing at all. In other words, they no longer contribute any overhead, since the statements are completely eliminated at pre-compile time. Therefore the assert() macros can stay permanently in the code as a form of documentation, imposing no overhead whatsoever.

5.13   WHITE-BOX TESTING

This simple, useful, and little-known macro can be used at many different levels of formality. Most programmers begin using it in an ad hoc manner, sprinkling them randomly through their code to verify that assumptions are being met, much as debugging statements are used.

---

[5] Since robust exception handling is not available in typical C environments, "raising an exception" is merely a euphemism for unconditionally exiting the program. In C environments that support full exception handling, assertionTrap() would actually raise an exception, Assertion Exception, which the programmer could intercept and possibly handle without unconditionally exiting the program.

Asserts can also be used in a more rigorous fashion called white-box testing. The assertions are written right into the body of a class, generally by the programmer who develops the class, exactly as before. The only difference is the degree of formality. The assertions state formal preconditions that must be met before a body of code will work properly. Preconditions verify that the initial conditions are as expected by the code; i.e., that the user provided the correct arguments and that the receiver's state is as required for the method to work correctly. Postconditions verify that the code produces the expected result. For example

```
@implementation Set:...
- add:anyObject {
        assert(isObject(anyObject))⁶;     // Precondition
        ... add the object
        assert([self contains:anyObject]);//Postcondition
        ...
}
```

Carefully chosen assertions are a particularly cogent and readable kind of documentation. Preconditions describe what the user must do before calling a method. Postconditions describe what the method will do as a result.

5.14   BLACK-BOX TESTING

White-box tests are written by the developer of the code. Although they are an effective and highly recommended documentation and debugging technique, they are not a formal specification/testing technique. This involves black-box, or external, tests that can be expressed by someone who is motivated to show that the code can be broken, not by the developer who is motivated to show that it works. The same macro is used for black-box testing, which differs from white-box testing in that the assertions are in a different file from the code being tested; in C, as a subroutine called a test procedure.

However, this separation reflects a fundamental distinction. With black-box testing, the component is no longer viewed from the inside, the viewpoint of its producer. Black-box tests view the component from the outside, the viewpoint of its consumer. Black-box testing is the first step across the boundary between implementing and specifying. Subsequent steps that

---

[6] isObject() is a predicate defined in Environment.m. It reports whether its argument is an instance of any class (e.g. whether the argument is certainly not a string or other pointer-based quantity, and is unlikely to be merely a large integer that happens to have the id of an object as its value).

transform this approach from a purely conventional testing technique familiar to any Q/A group, to a full-blown specification/testing technology that might transform software as John Hall's inspection gauges transformed manufacturing, will be outlined in subsequent sections.

As the following example should make clear, nothing very sophisticated is going on here, except for the change in viewpoint. This test procedure merely puts its argument, testedClass, to work by presenting it several inputs and examining its outputs by means of the same assertion macro described earlier. The return type of this subroutine is void because failed assertions are not reported by returning a boolean value, like FAIL, but by raising an AssertionFailed exception.

```
EXPORT void isaQueue(id testedClass)
{
        testedInstance = [testedClass new];
        qm1 = [QueueMember name:"qm1"];
        qm2 = [QueueMember name:"qm2"];
        qm3 = [QueueMember name:"qm3"];
        // Exercise an empty queue
        assert([testedInstance size] == 0);
        assert([testedInstance contains:qm1] == NO);
        // Build queue in reverse order
        [testedInstance enqueueHead:qm1];
        assert([testedInstance size] == 1);
        [testedInstance enqueueHead:qm2];
        assert([testedInstance size] == 2);
        [testedInstance enqueueHead:qm3];
        assert([testedInstance size] == 3);
        // Verify reverse ordered queue
        assert([testedInstance head] == qm3);
        assert([testedInstance tail] == qm1);
        assert([qm3 successor] == qm2);
        assert([qm2 successor] == qm1);
        assert([qm1 successor] == nil);
        assert([qm1 predecessor] == qm2);
        assert([qm2 predecessor] == qm3);
        assert([qm3 predecessor] == nil);
        // Remove central element
        obj = [testedInstance dequeue:qm2];
        assert(obj == qm2);
        assert([testedInstance size] == 2);
        assert([testedInstance contains:qm2] == NO);
        assert([testedInstance contains:qm1] == YES);
        assert([testedInstance contains:qm3] == YES);
```

```
        // Remove head of two-element queue
        obj = [testedInstance dequeueHead];
        assert(obj == qm3);
        assert([testedInstance size] == 1);
        // Remove tail of one-element queue
        obj = [testedInstance dequeueTail];
        assert(obj == qm1);
        assert([testedInstance size] == 0);
        // Rebuild the queue in reverse order
        [testedInstance enqueueTail:qm1];
        [testedInstance enqueueTail:qm2];
        [testedInstance enqueueTail:qm3];
        assert([testedInstance size] == 3);
        assert([testedInstance head] == qm1);
        assert([testedInstance tail] == qm3);
        assert([testedInstance dequeueTail] == qm3);
        assert([testedInstance dequeueTail] == qm2);
        assert([testedInstance dequeueTail] == qm1);
        assert([testedInstance size] == 0);
        return self;
}
```

These test procedures are like the tools used in mature engineering domains to verify that some object is within tolerance of its specification. A carpenter's square, for example, is separate from the boards to be tested and from the implementation tools used to create the boards. It can be applied to any board to verify that the board meets the specification of being square, within a tolerance implicit in the carpenter's square.

The loose tolerance of this test procedure is reflected in the fact that this test procedure doesn't exhaustively test every possible combination of inputs and use cases, since the cost of exhaustive testing quickly approaches infinity. If the cost of a failure is high, the higher price paid by the consumer would be reflected in a tighter-tolerance test procedure. The objective of this approach is not to prove that a component cannot fail. It is only to show that the component complies to its specification within tolerance, where tolerance is determined by how much the consumer is willing to pay for high-precision components.

5.15   BLACK-BOX SPECIFICATION/TESTING

One major problem with the simple black-box testing scheme is that classes are expected to exhibit common behaviors, but labor-intensive testing

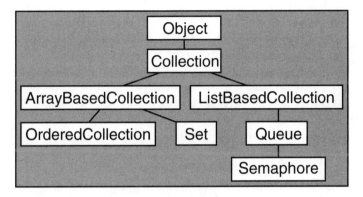

Figure 5.6 An implementation hierarchy describes how something in *implementation* is distinct from what it *is* or *does*.

schemes do not provide an automated scheme of ensuring that these common behaviors are actually present.

Figure 5.6 implies that Queue should exhibit some degree of commonality with ListBasedCollection, Collection, and Object. By policy, every class has a test procedure, and Queue is expected to pass, not only the isaQueue procedure shown above, but also the isaListBasedCollection, isaCollection, and isaObject test procedures of its superclasses.

The more advanced approach to be outlined below was motivated by the following inadequacies with conventional software testing practices. The flaw with the conventional scheme is that it assumes that the inheritance hierarchy reflects a specification of a component, whereas in practice the inheritance hierarchy quite often is used for implementation. Two clear examples of this can be seen in Fig. 5.6:

*OrderedCollection and Queue are intended to be independent implementations of the same specification. However, inheritance-based testing provides no way of ensuring that this is in fact the case by invoking the same test procedures on the two competing implementations, because they do not share a common superclass.

Since Semaphore was implemented as a subclass of Queue, Semaphore will be tested by test procedures that are inappropriate and that actually violate the semantics of Semaphore. Semaphores should support only wait and signal methods. Wait and signal do not behave properly, causing isaSemaphore() to fail, if arbitrary elements are added and removed by isaQueue(), which inappropriately invokes methods inherited from Queue.

The inheritance hierarchy supported by programming languages does not reflect a specification. Inheritance shows how the component was implanted, not what it *does*. The new approach takes issue with those who argue that inheritance 'should' be reserved for specification. Rather, it argues that two inheritance hierarchies are needed, one to express an internal makeup of a component, its implementation, and a separate hierarchy (Fig. 5.7) to express the external static and dynamic properties of a component, its specification. This new hierarchy, the specification hierarchy, guides the test procedure assembly process independently of the implementation hierarchy in Fig. 5.6. Now Queue and OrderedCollection can be tested by the same test procedure, isaIndexableCollection, and Semaphore can be tested by only isaSemaphore() and isaObject(), as required by the specification of Semaphore.

This new hierarchy is supported within a separate tool, a specification tool. This tool is strictly analogous to an object-oriented programming language, in that it also uses inheritance to classify things according to their similarities and to build things from libraries of reusable components. However, the 'things' that it builds are not implementations of some specification, but gauges that determine whether implementations comply to their specification within tolerance. The specification language constructs these gauges by merely assembling test procedures from a library according to the specification hierarchy as outlined above.

5.16   CLASSES AND INHERITANCE

The confusion of implementation and specification is particularly prominent in that most fashionable of object-oriented features: inheritance. As used in object-oriented language circles, inheritance is the Blanchard lathe of soft-

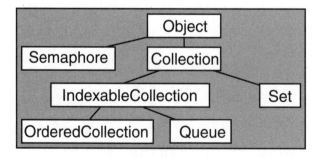

Figure 5.7 A specification hierarchy describes the *meaning* of the component with respect to the viewpoint of its user. Of course there can be, and generally are, many different users and therefore many such views. Compare the hierarchy shown here with the implementation hierarchy of Fig. 5.6.

ware: a powerful and important tool for creating new classes from existing ones, but not nearly as useful for specifying static properties such as how they fit into their environment, and useless for describing dynamic properties such as what these classes do.

Rather than laboriously building each new class by hand, inheritance copies functionality from a network of existing classes to create a new class that is, until the programmer begins overriding or adding methods, correct by construction. Such hierarchies show how the internals of a class were constructed. They say nothing (or worse, mislead) about the specification of the class, the static and dynamic properties that the class offers to its consumers.

For example, Fig. 5.6 shows the implementation hierarchy for a Semaphore class that inherits four existing classes. Contrast this implementation hierarchy with the following:

- Semaphores are a kind of Queue only from the arcane viewpoint of their author. This hierarchy resulted from a speed optimization of no interest to consumers, who should view them as scheduling primitives with only wait and signal methods. How will the producer tell the consumer to avoid some, many, or all of those irrelevant and dangerous methods being inherited from all four superclasses? And if such static issues are handled statically, where will dynamic ones be handled—the vital question of what Semaphore does?

- This hierarchy says quite explicitly that OrderedCollection is similar to Set and dissimilar to Queue. However, exactly the opposite is true. Queue is functionally identical to OrderedCollection. I carefully handcrafted Queue to have each of OrderedCollection's methods, each with precisely the same semantics, to show that encapsulation lets the internals of a class be revised without affecting its externals. But how will the consumer discover that OrderedCollection and Queue provide the same functionality, that they are competing implementations of precisely the same specification? And how will any commercially significant differences, like time/space trade-offs, be expressed independently of these similarities?

Manufacturing handles this issue by providing two separate classes of tools, implementation tools for the producer's side of the interface, and specification tools for the consumer's side. Shouldn't we do likewise? Shouldn't conceptual aids like inheritance be used on both sides, but separately, just as Blanchard's lathe and Hall's inspection gauges deal with different views of the same interface of the object? Shouldn't the consumer interface of Semaphore be expressed in an explicit specification hierarchy as in Fig. 5.7, independently of the producer's implementation hierarchy?

Just as a caliper is not a 'higher level' lathe, a specification tool is not a higher level implementation tool. Specification is not the job of the implementation tool. Separating the two would eliminate performance as a constraint on the specification tool, allowing knowledge representation shells that already support rich conceptual relationships to be used as basis for a specification language 'compiler'. The two tools can be deployed as coequal partners, both central to the development process, as in Fig. 5.8.

The primitives of the specification language are ordinary test procedures: predicates with a single argument that identifies the putative implementation to be tested. A test procedure exercises its argument to determine whether it behaves according to its specification. For example, a putative duck is an acceptable duck upon passing the isADuck gauge. The specification compiler builds this gauge by assembling walksLikeADuck and quacksLikeADuck test procedures from the library. The 'compiler' is simply an off-the-shelf knowledge representation tool for invoking test procedures stored for this purpose in the test procedure library.

The test procedure libraries play the role in software as elsewhere: defining the shared vocabulary that makes producer-consumer dialog possible. For example, in "I need a pound of roofing nails," pound is defined by a test procedure involving a scale, and nail by a test procedure involving shape recognition by the natural senses. Test procedures are particularly crucial in software because of the inability of the natural senses to contribute to the specification of otherwise intangible software products like Stack or Set.

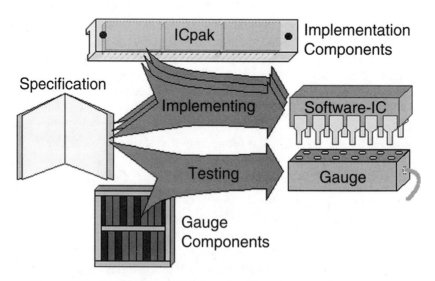

Figure 5.8 In mature engineering domains, the specification process has as much emphasis as the implementation process.

Making software tangible and observable, rather than intangible and speculative, is the first step to making software engineering and computer science a reality.

Test procedures collect 'operational', or indirect, measurements of what we'd really like to know, the quality of the product as perceived by the customer. They monitor the consumer's interface, rather than our traditional focus on the producer's interface (i.e., by counting lines of code, cyclomatic complexity, Halsted metrics, etc.). This knowledge of how product quality varies over time can then be fed back to improve the process through statistical quality control techniques, as described by W.E. Deming, that play such a key role in manufacturing today.

The novelty of this approach is threefold:

* It applies inheritance concepts not only to implementation, but to specification and testing, thus making the specification explicit.

* It preserves test procedures for reuse across different implementations, versions, or ports through an inheritance hierarchy.

* It distributes the specifications and test procedures between producers and consumers to define a common vocabulary that both parties can use for agreeing on software semantics.

The implications could be immense, once we adjust to the cultural changes that this implies, a shift in power away from those who produce the code to those who consume it, and from those who control the implementations to those who control the specifications.

- Specification/testing languages could lead to less reliance on source code, new ways of documenting code for reuse, and fundamentally new ideas for classifying large libraries of code so it can be located readily in reference manuals, component catalogs, and browsers.

- Specification/testing languages could free us from our preoccupation with standardized processes (programming languages) and our neglect of standardized products (software components). Producers would be freed to use whatever language is best for each task, knowing that the consumer will compile the specification to determine whether the result is as specified.

- Specification/testing languages can provide rigor to open universe situations when compile-time type checking is not viable. For example, in the set example described earlier, the implementation-oriented declaration AbstractArray* was too restrictive because sets should work for members that are not subclasses of AbstractArray. However, the anonymous type id is unnecessarily permissive because sets do impose a protocol requirement that you'd like to check before runtime. But because specification/testing tools can induce static meanings (isADuck) from dynamic behavior (quacksLikeADuck), why not feed this back to the programming language as implementation-independent type declarations? This amounts to a new notion of type that encompasses both the static and dynamic properties, rather than the static implementation-oriented meaning of today.

Chapter 6

Out of the Crisis

Sluggo: What are you looking for, Nancy?
Nancy: My purse! I lost it out in the yard.
Sluggo: So why are you looking here in the house?
Nancy: Because the light is much better in here.

— Comic Strip Characters

The software community loves programming languages and methodologies. Software development tools are interesting and fun, and everybody loves the web. Conferences, magazines, and books abound for anything having to do with computer technology, particularly for object technologies right now. The bandwagon is rolling and everyone is climbing aboard! The band is playing, the lights are blazing, and everyone has gathered to celebrate yet another latest and greatest technology. Who wants to discuss money with beancounters: managers, customers, accountants, executives, and venture capitalists? The cool people are all here at the party. And the light is much better in here.

I am as guilty of this as anybody. I spent my career building software development tools, believing that better software development processes, tools, languages, and methodologies, might get at the software crisis. My first book was about object-oriented programming languages and class libraries. I even cofounded a company, Stepstone, to build tools and objects for sale. This experience led me to realize that this focus on technology is treating the symptoms of the disease and not the disease itself. This grew into the realization that the software crisis does not originate where it can corrected by languages, tools, and methodologies. It originates from bad economics, not bad technology. It originates from the absence of a robust way of buying, selling, and owning easily copied information-age goods. This undercuts the economic system that we take for granted with tangible goods.

To put this change in viewpoint, this paradigm shift, in the starkest possible manner, imagine that we simply terminated all work on software development technology, canceled all conferences on software development tools, and even got rid of all tools more advanced than, say, assembly language.

Suppose that we fixed the economics instead. Suppose that we deployed a revenue collection technology, a way of robustly buying, selling, and own-

143

ing information-age goods throughout a multigranular structure of production. Suppose we made it possible for people to understand how they could earn a living by building components others might buy. If all this were done, wouldn't people overcome whatever obstacles stood in their path to satisfy the demand, *even with primitive tools such as assembly language*? And wouldn't suitable tools, tools even more advanced than we have today, soon follow?

6.1 OUT OF THE CRISIS

If you're primarily concerned with the here-and-now problems of buying and selling software today, this chapter is likely to seem irrelevant, if not repugnant, to you. I will spend very little time on dongles, or shareware, or demoware, or secure web servers, or license servers, or cryptography, or other unigranular solutions that are becoming commercially significant today. For these, I refer you to the trade press or to the following web pages in which I track these issues:

```
http://www.virtualschool.edu/bcox/ElectronicProperty.html
```

The problem we'll be addressing here won't seem relevant because software really isn't this way today. It is built as an artistic endeavor of fabricating software from first principles, the way San Ildefonso Pueblo Indians build pottery, not as an industrial activity that involves assembling components from markets as engineering disciplines build tangible goods.

Chapter 2 argued that the software crisis originates from the inability of the software industry to organize itself into an advanced social order. The tale of the wooden pencil showed that manufactured objects can be produced by people and companies linked by long-range economic forces arising from these exchange transactions. But since our goods are made of bits, traditional exchange transactions don't apply since bits can be replicated so readily. Therefore an elaborate structure of production hierarchies has never evolved for software. Regardless of the demand for the fruits of the potential structure of a production tree, which is clearly very high today, such trees cannot evolve so long as there's no way to pass energy down to its roots.

Figure 6.1 shows the human-centric system we'll be addressing as a simplified diagram of the sort that I introduced in Chapter 2. The circles represent economic actors and the lines represent the forces maintaining them in their role. Since traditional exchange transactions don't propagate incentives through electronic goods hierarchies, we'll be concentrating on new ones that can.

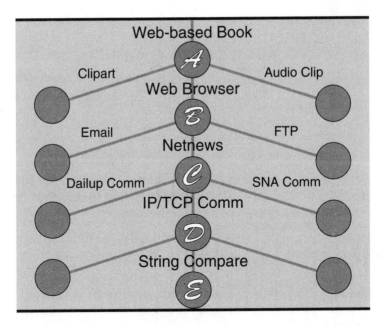

Figure 6.1 The circles represent economic actors, which might be either individuals or companies. The lines represent bidirectional exchange.

This figure intends to suggest a structure of production for an electronic property comparable to the one Chapter 2 detailed for wooden pencils. The author, Actor A, wrote most of the largest granularity object in this figure, an electronic book. But he used clip art, audio clips, and a web browser of the lower levels in this hierarchy: an artist, a programmer (Actor B), and a musician.

Notice that the web browser and the book are not treated as different. This is a departure from the established custom of carefully separating programs and data. We will be adopting the object-oriented view that everything in this figure is an object, and that every object is somebody's electronic property. These objects can be active (software) or passive (data) as they need to be to accomplish their function.

The programmer (Actor B) chose to concentrate his attention on building a good web browser. Therefore he relied on others for objects capable of supporting E-mail, netnews, and ftp features. The netnews programmer (Actor C) chose to support several different communication options (modem dialup, IP/TCP, and SNA), and acquired these from even lower-level sources. For completeness, Actor E provides the traditional example of an ultra-small-granularity object, the string compare subroutine, at the very bottom of this tree.

The decomposition in this figure poses a number of technical objections that I'll deal with later in this chapter:

- Why should browser owners let authors give away browsers with their books? How would the browser owners (or any of the subactors in this figure) get paid?

- Wouldn't the file size requirements become exhorbitant if every book contains not just the data but the programs for displaying it?

- Several of the objects in this figure could reuse the same subobjects. For example, the web browsing, E-mail, netnews, and ftp objects might need each of the communication options shown here.

- Aren't string compare routines too small to buy and sell as commercial properties? Wouldn't all programmers just build their own?

I want to defer discussion of these techno-centric issues until the human-centric issues arising from entrenched assumptions as to how electronic property can, or should be, bought and sold are firmly in mind. We automatically assume that each actor in Fig. 6.1 will purchase the products of lower-level actors to sell in as many copies of his own products as he might sell in the future.

We will be searching for ways that connect people into advanced social orders exactly like the social order that connects the farmers, millers, and bakers who provide us with bread. Just as millers don't license bakers the right to make as much flour as they might need to bake bread, we'll be departing from this assumption from the outset, regardless of the clear dominance of this assumption today.

6.2 COPYRIGHT AND USERIGHT

The Congress shall have power to promote the progress of science and useful arts, by securing for limited times to authors and inventors the exclusive right to their respective writings and discoveries.
— U.S. Constitution, Article I, Section 8, Clause 8

The music industry found themselves in our situation almost a century ago. The advent of broadcast technologies such as radio and television undercut the way musicians had earned their living in the past. One of the key innovations of this era was to recognize that copyright, the right to make replicas of music recordings, could be separated from useright, the right to use a recording however acquired, and that these two rights could be bundled, marketed, and sold in any combination the owner pleased.

In ancient times, before recording technologies were invented, there was no need to distinguish between copyright and useright. The only way to copy music was to round up the band for another live performance, and the only way to use it was to pull up another rock to the campfire.

When primitive recording technologies appeared with the introduction of music-writing notations, evolving to machine printing, it became possible to separate the production of music from its consumption. This made it possible to charge for these two rights independently. Musicians continued to collect revenues from ticket sales at the door, but they could also publish sheet music and derive revenue by selling sheet music copies.

Then music recording technologies such as piano rolls and phonograph records were introduced. How to measure the quality of a musician's effort now? Until then, quality was determined by the number of people a musician could draw and the price they would pay to hear; i.e., to use, an evening's performance. But now that a performer's value could be captured on a piano roll and paid for by the copy, would it be right for consumers to repeatedly receive this value for free through repeated play in their homes? Musicians worried that if they pursued this new possibility of per-copy revenue from piano roll sales, this would endanger per-use revenue from ticket sales at the door. After all, wouldn't people stop attending live performances if they could stay home and get the same entertainment from a player piano?

These questions were ultimately answered, by the musicians, the courts, and their listeners, on the basis of a simple common sense determinant that had nothing to do with morality, and everything to do with pragmatics. I call it the hassle factor. Since it was easy to obtain pay-per-use venues from ticket sales at the door, pay-per-use revenues were therefore collected. Since it was easy to obtain pay-per-copy revenues by selling piano rolls and phonograph records, pay-per-copy revenues were also collected. But it was a hassle to collect revenue for the use of copies once the customer took them home. So the hassle factor ruled and a potential source of revenue had to be foregone. The musicians realized that their revenues would never again be in direct proportion to the quality received by their listeners. They compensated as best they could by setting the price of piano roll copies high enough to compensate for the lack of a hassle-free way of collecting revenues each time a piano roll was used.

Keep this hassle factor in mind as we proceed through the following section. Also bear in mind Koji Kobayashi's vision of the thorough integration of computers, communications, and mankind, particularly in view of how this might redefine hassle factor with respect to computer software. For example, the hassle of paying for a piano roll is large since pianos are not connected to global telecommunications infrastructures. But the hassle factor will change radically when Kobayashi's vision begins to integrate computers and communications into the everyday affairs of mankind.

When electronic technologies like radio and television emerged on the scene, the music industry found itself in even deeper trouble. Broadcast technologies made it quite possible for a broadcaster to buy a single copy and replicate its value to millions of listeners. As you might well expect, monumental court battles ensued over this question, whose outcome can be stated quite simply. The courts ruled[1] that copyright law grants the musicians five different and distinct rights to their creations. The following table collects these five rights under the two headings that I used for this section. The first three rights pertain to **copyright**, the right to copy the work. The last two pertain to what I'll call **useright**, the right to use the intellectual property contained in a copy, regardless of how the copy was acquired.

Copyright

- The right to reproduce or copy the work.

- The right to distribute copies of the work to the public.

- The right to prepare derivative works.

Useright

- In the case of audio-visual works, the right to perform the work publicly.

- In the case of literary, musical, dramatic and choreographic works, pantomimes and pictorial, graphic or sculptural works, the right to display the work publicly.

The courts also affirmed that musicians can bundle these rights for sale in any manner they please. So when you purchase a recording at the music store, you're buying a specific bundle of rights. This bundle includes only (a) ownership of the physical copy and (b) a useright to use the music for personal enjoyment. It excludes other rights from the above list, such as the right to play the music over loudspeakers in bars and nightclubs, the right to play it over the airwaves, and so forth.

6.3 ASCAP AND BMI

ASCAP is the *American Society of Composers, Authors and Publishers,* a membership association of over 65,000 composers,

---

[1] *Text to Screen: Copyright Issues in the Electronic Age;* (c) 1989; Copyright Clearance Center, Inc.

songwriters, lyricists, and music publishers. The function of
ASCAP is to protect the rights of its members by licensing and
collecting royalties for the public performances of their
copyrighted works. These royalties are paid to members based
on surveys of performances of the works they wrote or
published.

—ASCAP Web; http://www.visualradio.com/ascap/as3a.htm

BMI (*Broadcast Musicians Incorporated*) is a nonprofit
organization representing more than 160,000 songwriters,
composers, and music publishers with a repertoire of more than
3,000,000 works in all areas of music. BMI distributes royalties to
its affiliates for the public performance and digital home
copying of their works, and in the future, will distribute
potential new royalties for the use of their works in cyberspace.

—BMI Web; http://bmi.com

Nightclubs, bars, and broadcasting stations who want these larger userights must purchase them separately. The publishers' first try was to insist
that each broadcasting station, bar, and nightclub negotiate and purchase
userights with each musician. And here the hassle factor emerged once
again. The buyers complained of the hassle of keeping precise records of
every airplay, and of the hassle of contacting and negotiating a deal with
each rights holder. As the demand for broadcast music grew, people began
to see the need for an industry-wide infrastructure, a revenue collection
scheme if you will, to simplify such negotiations for everyone concerned.

ASCAP was founded as a membership organization of music producers.
Its purpose was then, and is to this day, to reduce the hassle factor for pur-
chasing music userights and administering the revenue stream for its members. BMI is a similar organization that was originally founded to represent
the opposing interests of the broadcasting stations. Over time, the interests
of the two organizations converged to the point that I'll not further distinguish between them.

The innovation responsible for these organizations is called the blanket
license. The blanket license eliminates the requirement that useright customers keep detailed records and negotiate each use individually. A blanket
license conveys the right to use any and all of their member's music indis-
criminately. That is, it entirely does away with detailed per-play record-
keeping on the customer's part. The customer simply pays a prenegotiated
amount based on the number of listeners.

ASCAP funnels this revenue back to its members based on a formula that
recompenses each musician in proportion to the tune's popularity. This
involves measuring the number of times each musician's tunes are "used."

Although I've not seen this operation myself, I understand that it involves an army of workers who listen to recorded broadcasts all day, recording each airplay of each musician's tune on score sheets.

As you might well expect, neither the formula nor the sampling procedure is straightforward and both have been the subject of considerable wrangling in the courts. But a system nonetheless evolved that provides the musicians a way of packaging their product so that the interests of both producers and consumers are protected.

The courts have held that the producer owns all rights to his product, both copyrights and userights, and can bundle them however he pleases, subject only to the consumer's willingness to buy. The producer is free to sell records by the copy and to severely restrict userights. Ordinary shopping mall consumers accept this limitation, and most of us comply with the restriction. Producers accept that some amount of unauthorized usage will go on because the hassle factor of stopping it altogether is not worth the trouble. The producer is equally free to give the copies away and base revenue collection on usage, and broadcasting stations find that this rights bundle is more acceptable to them.

The music industry came to see that these two ways are not competitive with each another, but synergistic. It makes good business sense to sell records by the copy to those who agree to restrict their use of the music to personal entertainment. And it also makes sense to dispense with per-copy revenues for broadcasting studios by thrusting specially marked copies of exactly the same product on them for free. These customers pay no per-copy fees to acquire these records. They pay entirely in proportion to which tunes they play, how often, and to how large a potential audience.

6.4 ELECTRONIC GOODS

So what does all this have to with electronic objects? Are not radio broadcasting, bar, and nightclub analogies unlike the situation of software? After all, television and radio broadcast to many listeners, whereas computers narrowcast to individual computer users. Is not narrowcasting to broadcasting as signal is to noise?

There are indeed a number of differences of which these seem especially significant. I will deal with the first two first and reserve the third for an entire section unto itself.

- The first difference is a liability, something that is more difficult for software than for music. Music broadcasting is one-to-many, whereas software is one-to-one. The analog of the music broadcasting studio is the desktop computer, and the audience is the individual who uses that computer. This is a liability because it means that revenue collection must be done at the level of each individual software consumer instead of at the central broadcasting station.

- The second difference is an asset. Unlike software, music is passive, devoid of ability to monitor either how it is used or copied. Electronic objects are incapable of monitoring when they are copied, but software is uniquely capable of monitoring its use. Software is supremely capable of incrementing whatever usage counters you please, communicating this information through whatever communication channels you please, and of refusing to operate if the bills have not been paid.

- Finally, it is easier for most people to see computer software as a multigranular effort involving many people's property. We usually think of music as the creative product of an individual or band who fabricates it from first principles. Although digital music sampling technologies have embroiled the music industry in the multigranular property rights issue, too, this is rarer for music than for computer software and far less likely to succumb to technical solutions.

The courts held that the owner of an electronic property has, with limited exceptions, the right to bundle copyrights and userights to electronic properties in any combination he pleases and to sell them in any manner that the customers will pay to acquire. This is subject as always to the hassle factor and other laws of commercial enterprise, such as each party's ability to trust the other to buy and not steal.

With tangible goods, these laws are not solely the laws of man. Physical laws, such as conservation of mass, reduce the number of issues in which breakdowns in trust might occur. For example, the miller doesn't have to trust the baker not to replicate his own flour. They can build a stable and lasting relationship because of the baker's need to buy from the miller. With electronic goods, physical laws no longer prevent any customer from replicating as many components as he needs.

This question of trust is a crucial determinant of the market mechanism chosen for any particular circumstance. A strong safe and armed guards are appropriate for selling fine diamonds. But trust alone is usually enough for running a lemonade stand, particularly when backed by social law to handle the rare cases when trust alone might break down.

The first part of this chapter will concentrate on low-tech approaches to buying and selling goods that are composed trivially of bits. By low-tech, I mean approaches that work with computers and networks as they are today. Since today's computers and networks do not enforce ownership of multi-granular goods, these low-tech solutions rely almost exclusively on trust and social law.

Later, I'll show one way in which technology could be built that could play the role for commerce in electronic goods that conservation of mass plays for in tangible goods. The advantage of these higher-tech approaches is that this reduced the need to rely solely on trust and legal enforcement. This makes the effort of negotiating each transaction so high that the lower levels of Fig. 6.1 become ludicrously infeasible. The obvious disadvantage is that this involves additional technology, with all that this implies to the installed base of existing computers. The less obvious disadvantage is the paradigm shifts implied by adopting technology-based solutions to the ownership of electronic property.

6.5 LARGE-GRANULARITY SOLUTIONS

Copy protection amounts to a technical way of enforcing the first of the rights that the law grants to owners of information-age properties: the right to restrict copying. Copy protection prevents customers from acquiring a copy until a fee has been paid. The simplest example is that of physically controlling access to the bits, for example by locking up software in shrink-wrap boxes behind the counters of a software store.

Even though this approach supports a large and thriving industry today and so is clearly commercially significant, it does not address the problem we're addressing in this chapter. Although it works well for the large-granularity components at the top of Fig. 6.1, it is not an option for the small-granularity components at the root of this tree. Shelf space in software stores is so competitive that only large, relatively expensive components for a mass market can compete there.

A variation that might be more acceptable is to logically restrict access to the bits through encryption and similar means. For example, CD disks are now being distributed with a large inventory of software (or fonts, or clip art) stored on them. Several means, including encryption and disabling key features, block full access to the software until money has been paid for a key to unlock the full functionality of the software.

Shareware and its numerous variations takes a different approach that is similar in some ways to the one to be advocated in this book. Shareware products provide no revenue collection enforcement at all and rely (to varying degrees) on the honesty of customers. In the extreme form, the shareware author simply asks customers to send money if they find the product genuinely useful. Others use more-or-less gentle devices such as mildly annoying delays or splash screens to perpetually remind the user that paying the fee will remove the annoyance. Crippleware and demoware escalates the protection by disabling some key feature altogether, such as the Save feature in a word processor. Often the code is present but is prevented from working until you send money for an encryption key.

None of these low-tech schemes achieves the same results for smaller-granularity components as the pay-to-acquire basis of commerce in tangible goods. We continue to fabricate everything from first principles because there is no reliable way for those who might build smaller components to get paid. No one invests in fabricating small software for others to assemble because the low-tech revenue protection schemes do not work for small-granularity, low-priced objects such as reusable software components.

6.6 ROYALTIES

At least in the United States, copyright law estalishes two separate ways of bundling electronic property for sale: copyright and useright. These two rights can be bundled for sale in any combination. I treat them separately because useright-based approaches involve technology that is not readily available today.

Paradoxically, even though royalties are a copyright-based solution to small-granularity commerce that involves no technology that isn't widely available, royalties are almost never used today. Copyright is understood to mean that for a fee a customer has the right to replicate as many copies of a vendor's component as the customer might ever want. This understanding is implied in the industry-wide reluctance to even consider royalties as a payment option for software component suppliers.

For example, consider the lower levels of the hierarchy in Fig. 6.1. Vendor E's best current option is to attach his small-granularity product to something much larger, such as, in Stepstone's case, a compiler. The string compare component is then perceived as free by the customer and by the vendor as a cost center, not a profit center. This almost guarantees that managers and stockholders will see inadequate incentives to test, document, and maintain reusable components to the point that others will be prepared to reuse them.

The other option, which is somewhat less feasible, is to bundle the string compare routine with a large number of other components to produce a library large enough to be worth the trouble of marketing it. The accepted model for selling such libraries today is to charge a single relatively large fee, typically in the $500-$5000 range, for a license that allows the customer to include the library in larger applications.

This leads directly to the debilitating consequences I discussed in Chapter 2. Since the price is large and not proportional to utility, the vendor's income arrives all at once at the very beginning of the relationship with the customer. This leads to precisely the same dysfunction that farmers and millers would suffer if the miller sold the baker a license to replicate all the wheat and flour he might ever need in advance. Since the fee is large and fixed, small bakers couldn't afford it and large bakers would have an unfair advantage. Worse yet, the miller would have no incentive to improve the product over time.

Royalties solve this problem in a technology-free manner that could be adopted overnight. Instead of paying a large fee up-front, all customers, large and small, get the component for free. Later, when they begin to sell their own products based on this component, they pay a (negotiated) fee for using their subvendor's product. The subvendor now receives a continuing revenue stream that is directly proportional to the utility his component provides to his customers. Customers who sell more, pay more, exactly as in the miller/baker example.

For reasons I've never understood, the software industry is almost single-mindedly resistant to paying royalties. Part of the resistance is that since natural law (conservation of mass) can't be relied on to meter consumption of goods made of bits in other people's products, the subvendor must rely entirely on trust, backed by social law and negotiated contracts. The costs of negotiating and administrating such agreements, and the buyer's reluctance to expose himself to lawsuits if the trust relationship comes into dispute, no doubt contribute to this reluctance.

Nonetheless, I find the depth of this resistance so mystifying that I am unable to provide a better explanation than that this is a cultural taboo, analogous to the examples of other paradigmatic taboos scattered throughout this book.

6.7 USAGE-BASED SOLUTIONS

Because of this resistance to cooperative relationships that are based entirely on trust, the remainder of this book will concentrate on technical solutions. The emphasis of these solutions will be in using technology to compensate for the absence of conservation of mass for commerce in electronic goods. Userights are the second of the rights established by U.S. Copyright Law. This affirms the vendor's right to prevent users from using replicas of software, regardless of how they acquired them, until the vendor's fee has been paid.

The reason I'll be focusing so heavily on userights instead of copyrights in what follows is because of a simple commonsense observation. *Software is completely unable to determine that it has been copied but uniquely able to determine that it has been used.* The software vendor simply includes code in his product that counts how many times the software has been invoked.

These invocation counts are recorded in a tamper-resistant revenue collection infrastructure, where they become the basis for commercial terms and conditions, exactly as tangible terms and conditions are based on conservation of mass. This code is intrinsic in the software exactly as conservation of mass is intrinsic in tangible goods. This makes it possible to build extended structures of production hierarchies based on buying and selling other people's goods.

6.8 SUPERDISTRIBUTION

Experience in human society and abstract analysis in economics both indicate that market mechanisms and price systems can be surprisingly effective in coordinating actions in complex systems. They integrate knowledge from diverse sources; they are robust in the face of experimentation; they encourage cooperative relationships; and they are inherently parallel in operation. All these properties are of value not just in society, but in computational systems: markets are an abstraction that need not be limited to societies of talking primates.
— Miller and Drexler (1988b, p. 137), *Markets and Computation: Agoric Open Systems; The Ecology of Computation;* edited by B. A. Huberman; Amsterdam: North-Holland

Superdistribution is the name of one such approach. The name was coined by Dr. Ryoichi Mori in connection with work supported by JEIDA (Japan Electronics Industry Development Association), a Japanese industry-wide consortium. Mori has been developing this concept since 1987. However, similar approaches go by other names. My own early work refers to a similar idea by the term, CopyFree Software; i.e., software that can be freely copied and distributed without charge but with revenue collection based on usage. Ted Nelson discussed a similar approach in his early book, *Dream Machines,* in connection with incentive structures for hypertext documents. Mark Miller of Xanadu refers to a similar idea by the term 'pay-per-use software', a term that I avoid because this encourages the misunderstanding that invocation-based metering implies pay-to-use terms and conditions. The concept is closely related to usage-based revenue collection schemes in other industries, including telephony, music broadcasting, and cable video.

> *Superdistribution is an approach to distributing software in which software is made available freely and without restriction but is protected from modifications and modes of usage not authorized by its vendor. Superdistribution relies neither on law nor ethics to achieve these protections; instead it is achieved through a combination of electronic devices, software and administrative arrangements whose global design we call the "Superdistribution Architecture".*

> *The concept was invented by Mori in 1983. Since 1987, work on distribution has been carried out by a committee of the Japan Electronics Industry Development Association (JEIDA), a non-profit industry wide organization. That committee is now known as the Superdistribution Technology Research Committee.*
>
> — *Superdistribution: The Concept and the Architecture;*
> Ryoichi Mori and Masaji Kawahara[2]

Superdistribution uses technology in the revenue collection side of the equation to obtain a higher teeth-to-tail ratio than is possible through legal and moral sanctions alone. Its goal is to provide a meter that supports revenue collection for components of any granularity, and by extension, any other form of information that can be packaged within such software, as a commercially salable commodity. This allows electronic information to be treated as an asset, an information-age product that can be bought and sold much like the tangible products of the age of manufacturing.

Superdistribution does not apply to all kinds of information. It does not allow ephemeral assets such as wisdom to be packaged for sale. It applies only to electronic property. This encompasses a very large segment of the intellectual property problem, the very segment that most computer users are concerned with.

Superdistribution arises from the observation that a consumer acquires nothing of value when he acquires a piece of software. A copy costs nothing to make. And it confers on the recipient a liability, not an asset, for the copy occupies space that might be used for other purposes. The owner of the software doesn't lose anything of value when others acquire the bits. The owner loses money only if the value of the software is acquired by a consumer without similar value being received by the owner.

Superdistribution carries the one-to-many model of the broadcasting industry into the one-to-one relationship between personal computer and user.

---

[2] *Superdistribution: An Overview and the Current Status;* Ryoichi Mori and Masaji Kawahara; ISEC 89-44; and *Superdistribution: The Concept and the Architecture;* Ryoichi Mori and Masaji Kawahara; The Transactions of the IEICE Vol. E 73, No. 7, July 1990.

It provides revenue protection as robustly as with copy protection, but in a radically different manner that does not rely on copy protection at all. In fact, superdistribution is the opposite of copy protection, a way of locking up the bits. Superdistribution unlocks the bits by actively thrusting the bits into the user's hands in the hopes that they will be used. The revenue stream originates when the software is used, not when it is acquired. Copy protection compounds the problems arising from the intrinsic invisibility of software. Superdistribution actively addresses this problem by actively projecting the software into the hands of potential users, where it can to some extent serve as its own advertisement.

Superdistribution decouples revenue collection from software distribution. Revenue protection is not based on copy protection, but on the fact that copies have strings attached. These strings invoke a piece of technology that monitors usage of the software, whenever and wherever it is used. This software will not run except when this technology is present, responsive, and willing. Superdistibuted software will not operate on computers that do not have the superdistribution technology installed, and the technology will prevent the software from operating if the user fails to cooperate. For example, the software will not operate if the user refuses to allow the technology to upload usage information to an administrative organization for billing, if the user tampers with the technology, or if monthly usage bills are not paid.

This administrative organization is a kind of utility company whose product is information userights. The superdistribution technology is analogous to the meter that a utility company installs in one's house. The organization converts usage counts into a revenue stream by issuing a monthly usage bill to whoever used the software. The administrative organization then channels this revenue stream to its owners, the software developers and information providers, minus the cost of providing this administrative function.

This brief synopsis should suffice to explain what superdistribution is. But this explanation is certain to have raised a large number of objections. "This infringes on personal computing!" "What about information privacy?" "I'd rather own software than rent it!" "Software should be free!" "How would the superdistribution infrastructure come into existence?" And of course, "Software is too expensive already! Does this maniac expect me to pay even more?"

These are all valid issues and I'll address each of them in turn. None of them is particularly difficult compared to the massive chicken versus egg problems of getting large paradigm shifting innovations off the ground. But rather than address these issues now, I want to first describe the incentives for considering such a change. The next section will describe a particular scenario through which the change might unfold, and how the resulting system might operate in practice.

The market for computer hardware and software is still mostly in American hands. Assume that those who are now successful in this market continue focusing, much as they have on the past, on new features (graphics screens, then bigger screens, then color screens, now hypermedia, and so forth) that require customers to buy ever more powerful hardware. This is all quite understandable so long as selling hardware and software by the copy is the only way to make money. But superdistribution introduces other ways to make money. This is a unique opportunity for vendors who have never been particularly successful under the old paradigm to show their customers the advantages of changing the rules.

Insofar as I know, the system that I'll describe here is not a reality anywhere in the world, except as the vision behind the JEIDA initiative in Japan. The simplest and most forceful way of showing the implications of these initiatives is to pull out my crystal ball and look ahead a few years to describe one of several scenarios under which superdistribution might impinge upon the hardware and software market that is largely dominated by the United States today.

Superdistribution was brought to my attention by a pair of research papers from a Japanese industry-wide consortium called JEIDA (Japanese Electronics Industrial Development Association). The papers are primarily about a way in which VLSI silicon fabrication technologies could be exploited to provide tamper-resistant packaging of financially sensitive information. But their real significance is not these silicon-level details, but in the information that they're trying to protect. It is information about the usage of software on globally distributed personal computers, workstations, and mainframes. The tamper-resistant packaging is simply to prevent tampering with this information after it has been collected but before the information has been uploaded to an administrative organization for billing. The silicon-level implementation details are irrelevant to our purposes, and I'll not spend much time on them here. The significance is in the administrative system within which the superdistribution chip operates and in its implications to software revenues, to software quality, and to broader information-age issues in general.

Koji Kobayashi's vision of CC&M, the integration of computers and computers with mankind, is a foundation for the superdistribution approach. Superdistribution takes CC&M for granted. It assumes that a communications link is available to every computer user, and that this link can be used to upload software usage information to an administrative function. The precise nature of this link is not crucial. It is crucial only that there be one. Superdistribution does not use this link for high-bandwidth activities since superdistribution is not concerned with the distribution of software and information. Distribution is managed independently, through other conventional or unconventional means. Superdistribution uses a networking link to an administrative organization, but the link is only for reporting information on the usage of software for billing purposes. The bandwidth requirements for uploading this information are modest.

So to avoid the imponderables of when higher-speed telecommunications capabilities might be widely available, I'll postulate telecommunications capabilities no different than they are today. I will arbitrarily equate Kobayashi's vision with 14.4 kbps dialup communications over ordinary voice grade telephone lines. After all, most computers are already on desktops with a telephone right alongside. For older computers like mainframes, telephone access is not a major problem. The trend is toward laptops and palmtops, but these can be easily transported to telephone outlets occasionally. There are, of course, cases where access to a telephone cable would be a problem, but pay-by-copy software will always be an option for these. Pay-by-use software in a missile, for example, is inappropriate for obvious reasons. Let me also leave aside the computers that are already connected to higher tech local- or wide-area networks as well, since what I want to concentrate on here is not particularly dependent on the carrier technology.

If you prefer to assume that ISDN, or even better technologies, will be available soon to every computer owner, I'd be the last to disagree. Your crystal ball is as good as mine. Or if you believe that laptop and palmtop computers will mean that most computers will not be near a telephone outlet, that is fine too. The connectivity that superdistribution relies on does not need to be instantaneous connectivity. The information can be cached indefinitely within the computer and connected to a telephone only occasionally, say once a month.[3] Nor is telephone connectivity crucial to this idea. Electronic debit or credit card technology[4] could provide sufficient bandwidth by sending a pair of them back and forth each month in the mail.

Now imagine that hardware platforms, produced by the members of the JEIDA industry-wide consortium, start appearing on the market, exactly as Kobayashi said they should, with communications bundled right in. And imagine that each of these JEIDA computers contains a superdistribution chip built right into the heart of each system. Neither of these assumptions is particularly challenging, other than of course requiring that platform vendors cooperate. The price of 14.4 kbps modems is quite reasonable already. The cost of the chip and modem will be even less with the production

---

[3] There are, of course, boundary condition cases where pay-to-own will always be preferable , for controlling missile warheads, for example. Superdistribution is synergistic with pay-to-own just as renting is synergistic with owning.

[4] I am referring here to the flash card technology that is already beginning to compete with magnetic media such as floppy disks for some applications, such as in the palm-top computer industry.

volumes needed to put one into every JEIDA video game, PC, workstation, and mainframe.

In advance of the introduction comes marketing, carefully targeted at the interests of each major information-age interest group. The dominant message of this marketing is that, in this era of open systems and standards, JEIDA's systems are at least equal to what competing platforms are offering today. Quite possibly they'll be better in view of Japanese dominance of key hardware technologies such as disk drives, semiconductor memory, flat panel displays, and so forth. But let's put such speculations aside and assume that these platforms are no worse. They will run Windows 95, or MS/DOS, or Unix, or Macintosh, or VMS, or MVS acquired through financially rewarding licensing agreements with the owners of these proprietary software environments, as well as competing American platforms.

The ordinary pay-per-copy software that dominates the market today will continue to be available and will work fine on these platforms, both now and in the future. That is, superdistribution in no way inhibits the ability to buy software by the copy precisely as we do today. The superdistribution system pays absolutely no attention to ordinary software. It does not snoop on what the user is using a personal computer or workstation for. Superdistribution is completely passive until it is actively invoked by software specifically written to use its services. The chip is nothing more than an odometer for programs that are designed to use it as such. Its role is to create an additional way of distributing software in which revenue is collected independently of distribution. This new kind of software is specifically written to invoke the superdistribution chip. If you would rather own software you use heavily, you buy it just as you do today.

I will get to the superdistribution chip in a minute. Let us think about what that built-in modem could mean first. Built-in means that it is present in every JEIDA machine, from the smallest Nintendo or Sega game machine right up to the largest supercomputer. And Koji Kobayashi's vision of integrated computers and communications doesn't mean that platform vendors should leave communication services for an entirely different industry to screw up. This modem is merely the tangible evidence of an electronic communications infrastructure comparable in scope to, say, Compuserve, GEnie, or Internet. One might fervently hope that it would be better than these, but let's leave likely improvements out of this for now. Merely assume that the modem supports information services no better than Compuserve. The only improvement is that the network and the modem mean guaranteed connectivity with every user of a JEIDA computer, from game machines, to PCs, to workstations, to mainframes. Recall that Nintendo games are already in 25% of American households. This is nearly as common as telephone handsets, which is about as close as you can get to universal E-mail connectivity.

But so far I've not described anything that we couldn't do, that is, with some vision and cross-vendor cooperation. But throw in the paradigm shift stuff that the superdistribution system makes possible, and much larger possibilities begin to emerge. I will delay getting into how it all works later in this chapter. For now, let's concentrate on the *why*, as conveyed in marketing literature to carefully targeted information-age interest groups.

The large shrinkwrap application vendors, concerned about software piracy, would be shown how they can use the superdistribution system to sell software for these platforms exactly as they sell software today,[5] except that wholesale piracy is far more difficult than it is today.[6] I will show how superdistribution can be used to enforce pay-by-copy revenue collection later. For now just assume that the platform vendors would back up this assurance by giving preproduction machines to every major software vendor, and that the vendors would verify this claim thoroughly, undoubtedly by hiring hackers to attempt to crack the protection. Which system do you suppose they'd rather build software for? Conventional systems where piracy is an ever-present worry? Or JEIDA platforms with physical mechanisms that raise the piracy threshold substantially?

A slightly different picture is painted for those whose products are slowly changing high-value data, not procedure. This category includes such high-value products as font libraries and clip art libraries. This category could extend into parts of the newsletter and magazine markets, perhaps even to some portions of the dictionary, encyclopedia, and book markets. The marketing literature explains that the system provides no particular advantage for product distribution, apart from a 14.4 kbps modem on every platform. But the marketing message explains how to use superdistribution to decouple distribution and revenue collection. For example, it tells font and clip art vendors that decoupling means that they can now mail a disk with their entire product line to every computer user on their mailing list for free. Or they can make it available for free downloading over the network. But since revenue collection is independent of distribution, they'll receive every cent they're entitled to for each font or clip art drawing that the customer actually uses. The system, of course, is flexible enough to support the most creative marketing department penchant for mixing freebies and gotchas.

---

[5] This is not a conflict with my earlier assertion that ordinary pay-by-copy (i.e., piratable) software is not affected by the chip. If companies choose to distribute piratable software, it is no less piratable on these platforms. But superdistribution chips can also be used in a mode, if a company chooses to build software that uses this capability; that is nonpiratable. They can then sell that version of the software by the copy, just as they sell piratable software today.

[6] I don't claim that these defenses are impregnable, since this is always relative to the resources and determination of the attack. I only claim that the defenses are high enough to forestall wholesale copying across machines, not that an individual hacker cannot get around the defenses on a specific machine.

Superdistribution is of interest to many other groups as well, such as those who would like to provide information-age products ranging from reusable software components to news feeds to photographs. But let's set these groups aside for phase two. Merely consider how only these two innovations could be presented to end-users. JEIDA's machines would not be marketed as information plumbing technologies as the American platform vendors do today. They would be marketed as access to information itself, with information plumbing technologies discretely out of sight. The president of Stanley once encouraged his sales force to use the same approach in his famous instruction, "Sell holes, not drills."

The implications of this crucial difference would be hammered home through marketing. "These systems provide access to information itself. This is not yet another kind of irrelevant electronic plumbing." It is a way for dad to send E-mail from his workstation to junior at home on his video game, or in fact, to any user of a JEIDA platform. Here is a way to try out all of the software you've ever heard of, but with no heavy per-copy fee to be paid with no way of knowing whether you like the product. Here are all of the fonts and clip art that you've ever heard of. "Go ahead and try them out. You pay only for what you actually use."

Do I really need to spell out the consequences? As a software developer, whose systems would you build popular applications for? Wouldn't you prefer the nonpiratable JEIDA option? As a computer end-user, whose platform would you rather own? Given that both options do everything that you are accustomed to doing today, wouldn't you choose the one that allows you to use all the CopyFree software, font, and clip art libraries that keep turning up for free in your mailbox? Wouldn't you prefer to try out popular new software without the heavy upfront per-copy fee, paying only if you like it well enough to actually use it?

6.9 SUPERDISTRIBUTION AS A MARKET MECHANISM

Two extreme forms of organization are the command economy and the market economy. The former attempts to make economic trade-offs in a rational, centrally-formulated plan, and to implement that plan through detailed central direction of productive activity. The latter allows economic tradeoffs to be made to local decision makers, guided by price signals and constrained by general rules.

Should one expect markets to be applicable to processor time, memory space, and computational services inside computers? Steel mills, farms,

insurance companies, software firms...even vending machines...all
provide their goods and services in a market context; a mechanism that
spans so wide a range may well be stretched further.
— Miller and Drexler (1988b, p. 137), *Markets and Computation:*
Agoric Open Systems

Superdistribution is an information pump, a way of motivating people to supply information that other people are willing to buy. It addresses the key motivational question of "why." Why should I provide popular software on this platform and not that one? Why should I broadcast fonts and clip art to every customer on my mailing list, or upload them to the network? Why should I provide reusable software components, or newsfeeds, or photographs, for electronic distribution over the network? Superdistribution answers the key question, "What is in it for me?"

Another way of thinking about the significance of superdistribution is to think of it as an information market. A market is a vendor-neutral place, a location that anybody can visit to play the role of either buyer or seller. This might not be promoted in the initial marketing literature, but it is sure to be noticed eventually. The same motivational force that addresses the why question for font or newsfeed vendors is sure to be sensed by end-users. Some of these will have access to the kinds of locally specialized kinds of information that specialized niche markets would willingly buy. Where are the cheapest gasoline or grocery prices within five miles of your home? Among all of the spreadsheet packages on the network, which ones have the lowest usage fees? Some shoppers gladly stay up to date with such information. Others might willingly pay to acquire it.

The phase two possibilities are sure to be noticed by a world full of potential information-age entrepreneurs. Might it not occur to Aunt Nellie that her years of coupon clipping and comparison shopping has made her a world-class expert in grocery prices near her home? Might not information products by such entrepreneurs broaden the market in a way that high resolution color terminals and hypermedia could never do? Although the Aunt Nellies of this world are unlikely to care about information plumbing technologies, they certainly do care about information, and they certainly care about money. Might not her information product cause Uncle Tom and Aunt Martha and Cousin Kate to buy JEIDA machines merely to acquire access to Aunt Nellie's expertise? Would it not be much cheaper and easier for them to buy local price comparison information from Aunt Nellie than to go running around for it on their own? Would you not rather buy this kind of information from an expert like Aunt Nellie than to go floundering around for it yourself? Couldn't Aunt Nellie's new information product, and thousands of others like it, be the next generation killer app that the hardware community is always looking for, the software product that generates hardware sales on its own?

Superdistribution as an information market is much more than a metaphor. It means that all of the traditional market mechanisms continue to apply. For example, this is why I've never mentioned specific prices anywhere in this story. Markets are distributed decision-making machines, in the sense that it is the market that decides such matters as prices. Prices have two components. One is the monetary amount that a buyer pays to a seller, and is determined solely by market forces. The other is the transaction cost of getting in the car, driving to the store, finding a parking space, jostling with shoppers, and so forth; the costs of getting to the point of a buy/sell transaction. Superdistribution is concerned with only this latter aspect of pricing. It is solely a mechanism for reducing the transaction cost to a sufficiently low level that an information market becomes feasible.

6.10   SUPERDISTRIBUTION AND OBJECT TECHNOLOGIES

Before getting into how an invocation-based system might work in more detail, let me point out several things that it does not do. Superdistribution measures the usage of software on a user's computer in such a way that this usage can be bought and sold. That is, it makes it possible to assign a commercial value to software usage. It does not assign the value itself (the price). The price can be measured only by its effects on the buyer's mind. Superdistribution measures only usage of software in the buyer's computer.

For example, suppose that I have certain knowledge that a particular stock price will go up in value next month, and that I decide to sell this valuable information through an invocation-based system. This information could be represented quite concisely in an ordinary ASCII E-mail message. But superdistribution could not be used to sell information packaged as pure data. Superdistribution is solely concerned with usage of software, not usage of data.

To sell this information through superdistribution, passive data must first be encapsulated in an active carrier. This is where superdistribution ties in with object technologies. This carrier need be no more elaborate than a program that draws the information on the user's screen, just like the little programs with which some Macintosh documents are encapsulated today. But with valuable information such as this stock tip, this essential element might become quite elaborate. Much of the functionality of such an object would be accessible for no usage fee as advertising. Most of the program might be exclusively concerned with convincing the user of the value of the information that lies hidden, carefully encrypted within the sole method that incurs a usage fee.

Of course, this doesn't mean that you need to be a programmer to bring an information product to market. Generic information carrier application would soon appear, with convenient user interfaces to help nonprogrammers bring information products to market. Notice the specialization of labor that is already emerging right here. How does the information carrier's owner get paid when I sell one of my stock tip objects? It is actually quite straightforward. My customer will be billed for one usage of my stock tip object. This product is registered to me, so this money accrues to my account. However, my customer's usage record also reveals that my stock tip object has invoked an information carrier object that is owned by somebody else. The fee for this usage is a debit against my account and a credit to the vendor of the information carrier program.

This system of debits and credits continues recursively, just as in any market. It ripples right on down through whatever specialization of labor hierarchy contributed to bringing this vendor's product to market. For example, suppose that the information carrier application was constructed from reusable software components. These are registered by various third parties that my customer will never hear of. But these software components also invoke the superdistribution chip in my customer's computer so that each usage will appear in the usage record as a debit against the information carrier's owner and a credit to the owner of each of the components used to build it.

Chapter 7

Commerce Infrastructures

SOFTWARE DOLLARS SUCKED DOWN A BLACK HOLE:
Although U.S. corporations and institutions spend an annual $250
billion on software development, the Standish Group International
reports that only 16% of projects come in on budget, on time, with all
the features. Fifty-three percent are either over budget, delayed, have
fewer functions than planned, or any combination thereof.
                                        Investor's Business Daily 1/25/95

Today's applications are all *unigranular* goods. Software engineers' long-standing dream of reusable software components notwithstanding, software doesn't normally include other people's software inside it. This is best understood in contrast to the multigranular goods of everyday commerce in which goods are almost invariably produced by deep structure of production hierarchies involving everything from mines to refineries to factories. The software baker doesn't buy subcomponents from a market, but tills his own fields and grinds grain with his own metate just like the most primitive of agrarian tribes.

Commerce in tangible goods is based on *pay-to-acquire*. Payment is extracted upon acquisition of the goods. This system works well for tangible goods, which being made of atoms, abide by physical laws such as conservation of mass. Since the farmer and the miller know that each loaf sold by the baker is guaranteed by physical law to replenish the demand for wheat and flour, they have an incentive for cooperating with the baker by working to keep him stocked with raw materials.

This understanding breaks down for electronic goods made of bits. Unlike tangible goods, goods made of bits can be replicated by each customer whenever more is needed. There is no incentive for reusable software component developers comparable to the incentive of the farmer and the miller, so long as their customers can replicate new components each time they sell another copy of the goods that contain them.

From this human-centric observation springs the paradox that has confounded software engineering since its inception. During the very period in which hardware engineering is notorious for exponentially rising achievements, the software engineering community is best known for the software crisis.

This chapter describes a particular design for integrating computers, communications, and financial institutions into a system capable of supporting commerce in multigranular electronic goods. The goal is to show that the approach is feasible and to show how it might operate in practice. The solution described here involves a kind of jujitsu in that it turns the opponent's strengths into an advantage for the defense. The opponent in this case is the ease of replication of software.

The solution is to realize that the pay-to-acquire approach of the industrial age is not the only way of acquiring revenue, and that pay-to-acquire is both unnatural and infeasible for goods made bits. It is far more natural, and therefore more feasible. to build technical enforcement mechanisms that encourage people to acquire the bits freely and to derive revenue as the software is used. This would allow vendors to view uncontrolled copying as acquisition of new paying customers and not as lost sales.

7.1 COMMERCE INFRASTRUCTURES

This title originates from the analogy that I developed in the preface. The pony express was exclusively a *communication* infrastructure, similar to the electronic networks we have today. The railroads, by contrast, were a *commerce infrastructure*, capable of not only hauling invoices and letters, but also passengers, property, and freight.

The difference between railroads and the infrastructures we're concerned with here is twofold. First and most obviously, computer networks will never haul passengers, nor the other tangible goods of the industrial age. Second, and less obviously, railroads concentrated on what was then the hard part of industrial-age commerce, hauling goods made of atoms from one point to another. Since buying, selling, and owning goods was trivial for industrial-age goods, the railroads could ignore this problem and leave commerce to the merchants at each end of the line.

The commerce infrastructures we'll be concerned with here have exactly the opposite problem. Transportation of electronic goods is so easy that we needn't do anything further. Existing infrastructures such as internet, Compuserve, Prodigy, and America Online are already capable of transporting electronic goods at nearly the speed of light. The focus of our commerce infrastructure will be on the problem the railroads never had to address: providing a socially robust way of buying, selling, and owning multigranular goods made of bits.

The previous chapter introduced the overall approach under Ryoichi Mori's term, *Superdistribution*. He chose this name by analogy with superconductivity to suggest that information might flow freely, with zero resistance from copy protection and piracy.

However, in this chapter I'll leave that name behind and adopt the term *invocation-based revenue collection* instead. The first reason is that the design to be described here was arrived at independently of Dr. Mori's design by me in 1984, long before I'd heard of his work, while thinking about why the Stepstone Software-ICs had such radically different economics than the Intel Silicon-ICs.

I documented the design in a notarized patent workbook with the intention of filing a patent application. However, I ultimately decided that (a) a patent would inhibit the spread of this approach by reducing my credibility as its advocate, (b) the approach is a "way of doing business" which is specifically excluded from patent protection, and (c) I didn't wish to support or benefit from the devastation that the patent system has imposed on our society. History shows that ideas (inventions) are never original, but arise spontaneously, often in multiple places, when the climate is ready for them. It should not be possible to "own" such spontaneously arising notions. Likewise I didn't invent this idea. I learned it from the music industry which explored this trail more than a century ago.

The second reason for avoiding the term, superdistribution, is that it suggests that it has something to do with software distribution: But it actually has nothing to do with distribution at all. Distribution was an issue for industrial-age goods. It is hardly an issue for goods made of bits, goods that can be replicated and distributed at the speed of light. There is no distribution problem for software. There *is* a problem of collecting revenue for goods that can be distributed so readily.

Prodigy, America Online, Compuserve, and Internet all provide perfectly adequate distribution technologies, as do CD disks or floppy disks passed between friends. By disconnecting revenue collection from acquisition of bits, software can be freed to distribute, advertise, and collect revenue for *itself*.

This chapter will describe the revenue collection system shown in Fig. 7.1. The goal of this system is to move software from its primitive social organization of today, in which most members of the society fabricate everything from first principles, to an advanced social organization in which they assemble software from electronic goods produced by other members of a broad and deep structure of production. In other words, its goal is to support commerce in electronic goods that contain other peoples' goods, and who naturally wish to profit from the contribution they made to making the whole a success.

Revenue collection for tangible goods is based on physical laws such as conservation of mass. These laws ensure that the farmer and the miller profit in proportion to the baker's success by guaranteeing that each sale by the baker will renew his need to buy flour and thus wheat. Tangible goods inherit these laws from their origin in nature. Since electronic goods don't originate in nature but solely from the activities of people, a comparable basis for commerce can originate only from something we must build. That is, we didn't have to build a system comparable to the one described here to get conservation of mass to support commerce in newspapers, automobiles, and Twinkies.

7.2 USER INTERFACE

The system to be described in this chapter is shown in Fig. 7.1. At the right is a software user, who interacts with the system through the user interface to be described in this section. Users who only consume and never produce are rare in a multigranular structure of production, so the user interface equips every user with the infrastructure to play both roles.

The rest of the revenue collection infrastructure is shown at the top of this figure. Proceeding counter-clockwise, the infrastructure consists of low-level invocation-metering capabilities within each user's PC. This meter is low-

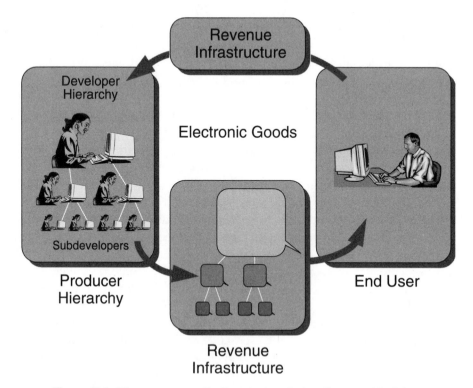

Figure 7.1 The revenue collection system to be discussed in this chapter.

level in exactly the sense that conservation of mass provides low-level infrastructure for commerce in tangible goods. It provides a tacit foundation upon which higher-level human understandings, such as commercial terms and conditions, can be based.

The remaining parts of the revenue infrastructure (top box in this figure) convey the invocation-metering outputs to a financial institution which manages financial accounts. Higher-level issues, such as converting invocation counts to financial amounts due and transferring those amounts between accounts, are handled exclusively at this level. The effect is that revenue flows from the end-user at the right and is distributed to each node of the structure of production that produced the goods the user used.

The remaining two blocks in this figure are to emphasize that the producer of goods is not just large corporations, and not just the developer at the top of the tree. Goods are produced by a tree of producers, each of whom must be incentivized fairly. The box at the bottom shows that the tree of producers, where each node in the tree assembles higher-level goods from components of subsuppliers, leads to a corresponding component hierarchy inside the goods they produce.

We will start our exploration of this revenue collection infrastructure from the outside, from the user interface. From the user interface, the system looks like a financial account management package, not unlike the ones that many of us use already to balance our checkbooks.

For those who are not familiar with such packages, Fig. 7.2 is a screen shot of Quicken, a representative of many such packages. The window shows a Quicken check register window into which I've typed several credits and debits:

This figure emphasizes account browsing functionality that is already available today. The following functions are not available in these packages today (Fig. 7.3):

Account Management: The dominant system today is for account data to be maintained centrally (by a bank) and for the user to type in data from monthly statements and check register receipts. Integration of computers, communications, and financial institutions implies that these functions would be done electronically. Later I'll show how the ability to obtain up to date financial balance information plays a crucial role in fraud prevention.

Product Registration: The user interface is fully equipped to allow the user to operate, not just as a consumer of other people's products, but to assemble other people's work into higher-level goods or to fabricate new ones from first principles. If registration involved manual intervention by other people (for example, via the telephone or paper documents), the central workload would be astronomical and introduce unnecessary opportunity for errors.

DATE	NUMBER	DESCRIPTION		DECREASE	✓	INCREASE		BALANCE	⇧
		MEMO	CATEGORY						
12/28	101	Opening Balance			✓	48	00	48	00
1994			[LectroNubbin]						
12/28	102	December Credits				19	40	67	40
1994									
12/28	103	December Debits		17	30			50	10
1994									
12/28									
1994									

LectroNubbin: Register

Save   Restore   SPLITS   Current Balance: $50.10

Figure 7.2 shows the credits and debits against an account that the user has named LectroNubbin. It shows the account balance history for the user's electronic property by that name. The decrease and increase columns show credits and debits against this account for the month of December. The credits arise from charges other users incurred by invoking this product. The debits are payments to other users whose electronic property is invoked by LectroNubbin.

User Registration: The user interface is an electronic product like any other and could be bought and sold similarly, by invocation or by acquisition. As the user interface changes hands, it carries with it the capability of registering its new owner electronically, again in order to reduce central workload and clerical errors.

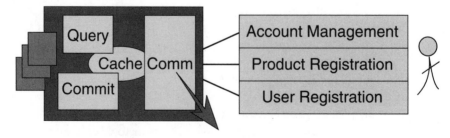

Figure 7.3 shows the top-level components of this system. The three components toward the right support the user interface. Those at the left are internal components that the user experiences only indirectly.

7.3 USER REGISTRATION

Existing account management packages automate only management of accounts, not their creation. You create a bank account by visiting the bank in person, or a credit card account by sending a form in the mail. Since these functions could be accomplished electronically, we assume for the sake of completeness that they will be. On the other hand, arguments for automating this user interface function are not strongly compelling except insofar as labor-saving advantages. Acquisition of an account is (in principle) a sufficiently rare event that we might as well drive to the bank in person.

The financial institution approves a new user registration just as it approves a credit card application. If the registration is accepted, an identifying number is issued that will be called the Customer ID, or *cid*, in what follows, even though this number actually identifies an *account* not a *person*.

One part of the user registration process may not be an obvious extension to past practices: The registration process is a process in which a legally binding contract is communicated and agreed to by two parties, the customer and the financial institution. The implications of this are best understood by comparison with commerce in tangible goods. When you buy a cabbage from the grocer, there is no need to register as a new customer, nor to sign a legally binding contract not to replicate that cabbage and give copies to all your friends. You don't need to agree not to depart from the implied terms and conditions of that sale because the terms and conditions are enforced by physical laws such as conservation of mass, not the laws of mankind.

However, the system to be described here involves technosocial law, as distinct from natural (self-enforcing) laws. Registration as a customer involves the acquisition of both rights and responsibilities. The customer acquires the right to use the system described here and simultaneously the responsibility to play by the rules. I will leave the precise rules to those trained in such matters. Customers would agree to such obvious matters as paying bills on time and refraining from various forms of tampering. The financial institution would make similar warranties with respect to paying their bills, refraining from fraud, and what use they can make of the invocation data this infrastructure would entrust to their keeping.

I am no lawyer. I have only been assured by those who are that such agreements are legally enforceable. Acquisition of an account involves establishing a contractual relationship with terms detailed during the registration process. As will become clearer once I describe how the technosocial system operates, this contractual law between individuals is based on the statutory laws of government, which are in turn rooted in constitutional copyright law. Although the enforceability of such contracts will no doubt be tested in court, it would be very surprising if agreements made during the registration process would not be binding in court.

The biggest single obstacle to this system being accepted and used is its fearsome information privacy implications. No company would use this system if it was thought that detailed information about exactly what software its employees are using this week could be made available to the highest bidder. No individual would use such a system if information about tastes in computer games or images, to pick only two issues of many, might be accessed by marketeers, detectives, and reporters in perpetuity.

Today we rely on natural law to limit such fears. We understand that it is not technically feasible to monitor what software we are using (although NSA's resistance to wider use of Tempest technology to limit electromagnetic emissions from computers suggests that the difficulty of such snooping is not as absolute as most people believe). We make a similar compromise each time we give up the almost absolute privacy of cash transactions and use a credit card instead.

7.4 INVOCATION METERING

The product name in Fig. 7.2, "LectroNubbin", was chosen to leave its actual functionality and granularity ambiguous. LectroNubbin might be a large-granularity object such as the computer applications the software market provides today. But such objects are not the best example to have in mind because simpler infrastructures are perfectly capable of supporting these already. A shrinkwrapped box in a software store handles revenue collection by acquisition quite nicely, as do CD disks whose contents can be enabled with a telephone call. So do the alternatives now becomming available, such as worldwide web servers that distribute software over networks only after access fees have been paid.

LectroNubbin might be a small component that can't stand on its own, such as a window object. Depending primarily on technical limitations of a deployed system, it could in principle be as small as a string compare subroutine. It could be a large component with enough functionality that ordinary people can understand and use it right out of the box. It might be a piece of clip art of a plate of steaming cookies, or a cookie recipe that this picture was purchased to illustrate. It might be an electronic cookbook that gathers recipes from many independent sources, or an electronic bookstore that collects such books for resale, or an electronic mall object with many such store objects inside it.

Exactly as with the structure of production that supplies tangible goods, the tree extends indefinitely. LectroNubbin is simply a name for the product at any of the nodes in this tree. It might be any electronic object that one person might produce to be bought by another. Or it might be the purchaser's higher-level goods, which encorporated LectroNubbin even *without its owner's knowledge or consent!*

LectroNubbin might as easily be network-oriented objects whose commercial value is far more obvious than these. It might be a carefully pruned and selected list of E-mail addresses to help charitable institutions generate mailings to funding institutions. Or lists of amateur bluegrass pickers in your neighborhood, which might be of interest to those hoping to jam or to sell guitars, picks, and strings to them. Or lists of those who have bought a home recently or had a new child, two examples with far more unsavory connotations to those who have been through the junk mail triggered by such experiences. The examples go on forever. Everybody knows something that somebody else might want to know bad enough to pay for it, and the reasons for wanting to know it span the entire range of human diversity.

7.5 QUERY AND COMMIT

Each of the electronic products in this deeply nested hierarchy of products contains the following pair of instructions:

```
meter.query(pid, ...) { /* OK to invoke? */
       ...              /* Value-bearing code */
meter.commit(pid, ...)  /* Commit this invocation */
```

Value-bearing code is enclosed within calls to a pair of instructions, *query* and *commit*. Stripped to their barest essentials (that is, ignoring tamper-proofing for now), these instructions are simple. They merely record the fact that they've been called, along with their arguments, in a tamper-resistant repository called the *cache*. Since products are nested within one another, records generated by the query/commit calls will be similarly nested in the cache.

The two instructions take the same argument list. The first argument is the *Product ID*, or *pid*. The pid identifies the value-bearing code contained within the query/commit instructions. The pid is obtained during the product registration process and is hard-coded into the two instructions before the product is delivered to the market. The elipsis means that both instructions accept variable-length argument lists. The extra arguments can be used by vendors who want to support specialized terms/conditions. For example, charging based on time or duration of use would be accomplished by providing the time as an additional argument. The argument list does not identify the person or product who is using this product because this is implicit in the enclosing record in the cache.

Periodically (say monthly), or when the cache level reaches a preset threshold, the cache initiates communication with the financial institution. This involves encrypting the cache contents and uploading them to be processed for billing. If the financial institution is satisfied with the status of this customer's account, it resets the cache and sets a flag that will allow the monthly cycle to continue. Otherwise this flag will prevent further use of invocation-based software until the account problem has been resolved.

7.6 WHO PAYS THE TAB?

The account that will ultimately be charged for invocations at each level is implicit in the hierarchy of nested records. The owner of each product is responsible for charges for subcomponents invoked by that person's property. This person is determinable from the enclosing record in the cache.

The outermost record in the cache is generated by the user's operating system. It creates the outermost record by simply calling query with the user's id in the place of the usual product id argument. Thereafter applications will be charged to the user, application subcomponents to the application, and so forth, all according to the nested relationship between query/commit records in the cache. For example, consider end-user A, who uses product B, which invokes subcomponent C.

This raises practical questions such as, "What about PCs that are shared with other users?" and "What about use of applications by employees on be-

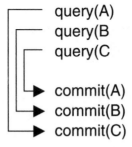

Figure 7.4 Charges for the use of component B are incurred by the owner of B's enclosing query, user A. Charges for component C will be incurred by C's enclosing query, the owner of component B. The outermost component, A in this case, is the user's operating system, which is treated for the purposes of charging as 'belonging' to the end-user.

half of their company?" The fundamental question here, "Who pays this tab?", boils down to one of, "Which account ID should be used for that crucial outermost record in the cache?" The answer ranges from technically easy but notoriously inadequate account identification devices such as passwords, to technically ambitious solutions such as smartcards and a variety of other devices on the market that a user can use to identify an account.

The range of feasible account identification devices, and broader ranging questions of how such devices are acquired, used, and protected in practice, would take us far beyond the scope of this book. However this account is identified, the issue that does concern us here is the question of how a cache-based system such as this would handle such human foibles as users who don't pay their bills.

The function of preventing use by invalid accounts, or further use by accounts in arrears, is handled by the query instruction. The query instruction checks account status by checking the flag that is set during the monthly negotiation with the financial institution. If the account number proposed in the query argument is not acceptable, the query instruction refuses to allow this user to use value-bearing software further, for example by raising a window to the effect that further access will not be allowed until the account is within threshold.

The query instruction tests this flag only for outermost query calls. This is because the user has a commercial relationship only with those whose components he or she uses directly. They have no relationship with, or awareness of, those whose components are used by that vendor. Outermost users must not be inconvenienced, or even made aware, that the vendor of some component they have no direct relationship with has not been paying his or her bills. That vendor's electronic property remains on the market and continues to function exactly as always. That is, the component continues charging its consumers and incurring charges for the components it uses exactly as always. That is, the relationship between software already on the market and the account is not subject to breakdowns in the relationship between the *persons* with an interest in that account, namely its owner and the financial institution that manages it. Disputes at this level are resolved by the legal contract established when the account was created, and are beyond the scope of the invocation-recording infrastructure described here.

7.7 TERMS, CONDITIONS, AND PRICES

Notice that prices have never entered the discussion. This is because the meters' query and commit instructions are totally unconcerned with prices. They record only that they've been called, for which products, and by whom. This is not an accidental oversight; it's a crucial feature of this design. The parts of the system discussed thus far play a role analogous to that which conservation of mass plays for tangible goods. Just as conservation of mass enforces the pay-to-acquire basis of commerce, the invocation-counting machinery described above merely collects information about invocations. Pricing, a subset of the more general topic of terms and conditions, enters at a much higher level.

This may be clearer with an example. Consider how the owner of an electronic cookbook might offer such an object for sale. Or more to the point, consider how you might react, as the customer, to various terms and conditions that might be offered to you. Both buyer and seller know that conventional paper-based cookbooks are readily available in conventional bookstores for, say, $20, under conventional pay-to-acquire terms and conditions. But you both know that the desire to own cookbooks in particular (in fact, books in general) is itself somewhat paradoxical in that the fate of most books is to be read once and then gather dust on a shelf. Cookbooks are even more special. Most customers might skim them cover to cover briefly to mark the recipes that interest us, then return time and again to the best ones but ignore the others.

The seller of every such property goes through a complex chain of reasoning to decide what pricing structure might maximize the attractiveness of this product with respect to its competition, knowing that the customer will do likewise. The outcome of the provider's guessing game is the terms and conditions under which the product will be tentatively introduced to the market.

So how might the cookbook be priced? It is really not our business to say. Prices are entirely up to its owner to propose and for the customer to dispose. Which is to say, prices are always set by markets in action, not by either side of the transaction and certainly not by external parties like me. This is why prices in particular, and terms and conditions in general, are not dealt with at the lowest levels of this system. Just as conservation of mass provides a basis for tangible commerce without concern for prices and terms/conditions, the same is true of the invocation-counting levels of this design. The PC-resident parts of this system are as simple and general as can possibly be arranged. The complicated matters are handled at the higher level and will be described in the section on *Terms and Conditions*.

7.8 TAMPER-RESISTANCE

I chose the term "tamper-resistance" instead of "tamper-proofing" for a reason. There is no such thing as tamper-proof, just various degrees of tamper-resistance. Tanks were invented during World War I to protect soldiers against enemy bullets and shrapnel. But the subsequent race between tank armor and anti-tank missile technology has shown that there is no iron-clad proof against whatever threat an enemy might mount, just differing degrees of resistance *to the level of threat expected.* As the eternal race between bank vaults and safe-cracking technology also attests, greater resistance always implies greater cost. Both examples also show that decreasing the threat can often accomplish the same ends more cheaply than increasing the resistance through brute force. We see this in operation every time we visit a shopping mall, where the technical barriers to shoplifting are almost as negligible as the barriers to software piracy in computers.

An exploration of the whole continuum of possibilities, between high resistance/high threat on the one hand and low resistance/low threat on the other, would take us far away from the main topic of this book. So rather than explore the whole continuum, I'll only demonstrate that the same continuum of threat/defense tradeoffs exists with respect to electronic goods that we take for granted in our everyday dealings with those who might not deserve our trust. I'll do this by characterizing two extrema in the continuum of threat/response possibilities. The argument is that if each of the extreme cases can be addressed, intermediate points on the continuum surely can, too.

First, it is useful to distinguish between protection of *goods* from protection of the *infrastructure* as a whole. The goods in this case are computer software. The query/commit calls are at the mercy of a dishonest user's debugger, barring draconian use of encryption or other technologies I can't easily imagine. Although vendors will be prone to some revenue loss through attacks on the products themselves, the loss is confined to that single user (and a close circle of dishonest friends) by laws already on the books against redistribution of commercial software, particularly software that has been modified for obviously dishonest purposes. Given that the entire software market operates today without the slightest technical barriers to copying, vendors are likely to accept this approach as a definite improvement.

Whereas cracking a product provides free access only to that product, cracking the infrastructure provides free access to *all* products. If the modification can be widely distributed over networks, the free access would extend to all *users.* Protection of the infrastructure is obviously crucial, so I'll concentrate on that here.

The strongest form of tamper-resistance (apart from encryption, which I'll discuss separately) involves rigorously eliminating everything from the infrastructure that a fraud-prone (or accident-prone) user might tamper with, particularly software that users could change with debuggers.

The good news is that this is possible. The query/commit instructions are extremely low-level, small (except for the cache), uncomplicated, and fast. They could easily be provided as silicon. In the short run they could be provided as add-on devices. This is only a short-term workaround due to cost and performance implications, but especially since any chip sockets, buses, cards, and cables create opportunities for the dishonest to substitute whatever they please if each interface is not protected in some manner. The ideal solution is for silicon vendors to add the two instructions into the basic instruction set of new CPU chips, where they'd receive the same protection that ADD and JSR instructions enjoy today.

The bad news is that this involves new hardware. This is infinitely more difficult than an infrastructure based entirely on software. However, a seldom-recognized implication of the shrinking prices of hardware points to a possible solution. Shrinking prices have shaved hardware vendors' profit margins razor thin. Peter Sprague founded Wave Systems around the insight he gained as Chairman of the Board of National Semiconductor Corporation. He realized that a computer manufacturer's profit on a $10,000 computer sale is only about $30 or so, and the manufacturer receives this lump-sum payment only when new computers are purchased. He reasoned that hardware manufacturers would be amenable to providing hardware support for revenue collection by allowing them to share in the take. He brought Wave Systems stock public via NASDAC in the fall of 1994, based on a silicon chip that he designed to support this business model. Although the Wave chip has some capabilities similar to those described here, it is designed to support sales of unigranular *data* (television broadcasts, stock feeds, etc.), not the multigranular *objects* (data encapsulated in software) addressed in this chapter.[1]

The opposite approach is comparable to that of department stores, whose tantalizing displays of merchandise provide few technical obstacles to shoplifting. Such stores base their prosperity largely on socially constructed barriers such as trust for the honest and prosecution for the dishonest minority. Such an infrastructure could be implemented entirely in software. The question is whether viable markets exist in which socially constructed barriers to cheating are sufficiently strong that technically constructed barriers can be as weak as leaving the revenue infrastructure entirely subject to tampering with ordinary debuggers.

---

[1] The Wave chip provides two distinct capabilities: fast (100mB/sec) decryption and what Sprague describes as 'dongle-like' capabilities. The encryption is not public key, but is based on the older DES (Data Encryption Standard), a widely accepted secret key standard. The biggest difference from the approach described here is that Sprague's thinking is oriented toward the industry as it is defined today; i.e., where products are large and for the most part unigranular, and vendors are large corporations, not individual citizens. The chip does not support the strong separation between invocation metering and computation of terms and conditions discussed in this chapter, but restricts vendors to the palette of terms/conditions supported by the chip.

Surprisingly, such markets not only exist but are potentially quite large. They exist in any large corporation that provides software for its employees to use. The dominant system in such corporations today is to simply purchase truckloads of shrinkwrapped software, manuals and all, and hand them out to their employees. This is unnecessarily expensive, both because of all that paper, but also because some employees don't need or use the software, or neglect to keep it up to date with new releases. Furthermore, such companies are liable if employees violate the terms and conditions on such sales, for example by installing copies on home computers for those agreements that don't allow this. These expensive and legal liability issues are now being addressed, in a unigranular sense, by license server technologies. These allow companies to install a single copy of software on a local area network where employees can use it. They also purchase a certain number of licenses which allow only that number of employees to use the software simultaneously. The same approach also prevents the software from working except on computers connected to the license server, for example at home.

License server technology addresses the stated problems with today's unigranular products quite neatly. The shortcoming is that they don't support multigranular products. However, the example does show that commercially significant environments exist in which a trust-intensive, software-based approach could be viable that does support multigranular commerce. License servers are just software and are equally prone to attack by debuggers, but they are rarely attacked because corporations have too much to lose from lawsuits to risk it. Individual employees don't have an incentive to tamper since the employer pays the bills.

Putting the infrastructure entirely in silicon versus entirely in software are two extremes on a continuum of ways to protect a revenue stream via technical versus social means. There is a large number of intermediate possibilities, each with different tradeoffs. It would take us too far afield to explore them all here, particularly since the techno-centric issues of this system are such a small part of the human-centric issues I'm trying to raise. Since the silicon-based option could support the broadest market, I'll assume this option in what follows.

7.9 TERMS AND CONDITIONS

The user interface mentioned earlier is the interface only to the user's account. Each electronic property provides its own. In the case of a small-granularity product such as a subroutine or a class, this might be as simple as the traditional README file. For an object large enough to run on its own, an electronic cookbook for example, the book might use more technically so-phisticated approaches. Some objects, an electronic bookstore for example, might be primarily concerned with attractively displaying information about the goods on its shelves, with the parts concerned with payment tucked away discreetly...but never, ever forgotten.

Stated more generally, each object has a free part and one or more value-bearing parts. The latter parts are each bracketed by the query/commit instructions that convey information about who has used them to a financial institution for billing. This raises the questions to be addressed in this section: "How does the billing institution know how much to charge?" and "How does the customer know what the billing institution will actually charge?"

The answer to the second question is just as for tangible goods. The price (and more generally, the terms and conditions) must be made available to the customer before purchase. Along with advertising, displaying the price, the terms, and conditions of the sale, is accomplished by the free part of an object. If the object is too small to stand on its own, this might be an ordinary README file. If the object is more capable, a cookbook, for example, its au-thor would state the terms and conditions as part of the advertising function of the free part of the object.

The first question involves a separate mechanism. The life of a product begins when its product identifier is issued by the financial institution. This is the concluding step of social agreement between the financial institution and the owner of the product. The institution agrees to collect revenue for the owner of this product in return for certain reciprocal agreements by the owner of the product. One of the items provided during this process is an al-gorithm that the financial institution will use during its billing cycle to con-vert invocation counts to financial amounts due. I will call these the *Terms and Conditions Algorithms*, or *TCAs* for short.

The role of each TCA is to examine a stream of query/commit records. Al-gorithms are extremely flexible tools, capable of implementing even the most complex terms from a stream of invocation records. For example, it is easy for such algorithms to implement the pay-to-own terms and conditions that dominate the market today by charging the full amount due on the first use by this customer and nothing thereafter. It is nearly as straightforward to implement pay-if-you-like-it by deferring the fee until the nth use. And, of course, radical incremental charging algorithms in which a much smaller fee is charged upon each use are even simpler.

Since there is exactly one TCA per product id, the product id (pid) identi-fies a set of terms of conditions and thereby the product itself. In this sense, product id is really a misnomer. For example, suppose that the owner of Lec-troNubbin decides to offer LectroNubbin for sale both under conventional pay-to-own conditions and the other charged proportionally to use. The base functionality of LectroNubbin might be exactly the same in the two versions. However, its free parts would be different and so would the product IDs of the two versions.

7.10 RULES OF FAIR TRADE

Owners are guided through the process of acquiring new product IDs (pids) by the product registration user interface. This process is rather involved because the registration process involves both human-centric and techno-centric issues. It is not as simple as just computing and issuing a unique number to anyone who asks. The biggest issue arises not from techno-centric issues, but the biggest human-centric issue of all…fraud. Everything is based on the simple invocation-metering logic discussed earlier. This logic is entirely mechanical and not the least smart about human frailties. These span the whole range from outright chicanery to unsophisticated producers who misunderstand the instructions and do things that could expose themselves to legal and financial liability.

The product registration process establishes a legally binding agreement between the owner of a product and a financial institution. The financial institution agrees to handle revenue collection for the owner and disbursement to the owner's subcontractors (owners of internal property), in return for a fee. The property owner agrees to engage in fair trade. The product registration process involves obtaining from the owner a legally binding warranty that the owner's agents, the property itself, and the terms/conditions algorithm engage in rules of fair trade such as these:

- Fairness to producers: Usage of subsupplier's property by the owners' property must be accurately accounted and paid.

- Fairness to consumers: Prices (and terms and conditions in general) must be accurately conveyed to the purchaser before any purchase is committed.

I will leave the precise legal terminology here to the lawyers and concentrate on how this doctrine of fairness relates to the metering infrastructure described earlier.

Notice that, assuming that the invocation recording infrastructure is truly tamper-proof and the query/commit calls are used as described earlier, fairness to producers, as far as invocation metering is concerned, is automatic.

Each query/commit pair in the cache identifies a product and thereby the person responsible for its charges. The enclosing pair identifies the product, and thus the person, to be charged for this invocation of this subcomponent.

The fairness to consumers issue arises from the fact that there is only a legal requirement, not a technical one, that the price reported to the consumer by the free part of an object is the same as the price actually charged. The price actually charged will be computed by the owner's Terms and Conditions Algorithm (TCA) from the invocation record. This computation occurs on the computer at the financial institution, not on the user's PC.

The risks here are analogous to those warehouse stores where the price charged at the checkout stand is not based on the price printed on a sticker but is looked up during checkout according to the product code. The technology is wide open to abuse by store owners, who can easily understate the price on the shelf in the hope that some customers won't notice that they were actually charged more during checkout.

Paradoxically this kind of unfairness is far more likely in stores. Since disagreements are handled by individual customers, the unscrupulous owner can count on some of them not to bother. The crucial difference is that the financial institution is a third party that holds a binding warranty that the store owner agrees not to engage in such practices. The evidence of malfeasance is fully accessible to all parties, who only need to download the same software from any network and compare the price in its free part with that actually charged in a trial transaction. The price charged is routinely accessible to every customer via the account balance interface. Although only a few might check, it takes only one to get the owner into deep water with all customers.

This kind of fraud is less likely than in warehouse stores because every customer has the ability to issue an unfair trade complaint against any unscrupulous owner, and to back that complaint with indisputable evidence. The financial institution is legally bound, by its agreement with all users, to investigate the charges and to apply appropriate sanctions for just complaints.

Although the financial institution might well call in external law enforcement for particularly egregious cases, it has powerful enforcement mechanisms of its own. For example, it could punish owners found guilty of breach of the fairness contract by refusing to charge for products that don't comply with the fairness doctrine. This could have extremely serious consequences since the owner's product has been broadcast over networks and exists on thousands of people's hard drives. His obligation to pay subcomponent vendors invoked by his product remains fully in force. Unscrupulous behavior on the revenue side is inhibited by the knowledge that this risks termination of revenue without forgiving obligations on the cost side of his ledger.

7.11 REGISTERING A PRODUCT

It should be clear that registering a product is not a matter to be undertaken lightly. Legally binding agreements are involved that can't be waved away lightly. Although the ultimate outcome of the process involves no more than issuing a unique number, the prior steps in the process are needed to ensure that everyone understands the agreement and takes action to ensure that the products they're responsible for abide by the agreements.

Two extremes on the skill continuum should be noticed here. The simplest case is the technological novice, such the stereotypical Aunt Nellie who wants to publish cookie recipes over the network. It is simpler because Aunt Nellie is no programmer. She would never encounter such low-level matters as the query/commit infrastructure directly. She would concentrate on cooking and leave software development to others. She would adopt somebody else's product presentation object from the net and encapsulate her cookie expertise inside that, such as, for example, ShrinkWrapIt, a commercial product developed by a hypothetical programmer, Jill, for sale to non-computer-specialists like Aunt Nellie.

Please bear in mind that I'm not using the cookie recipe and shrink wrap examples because I think cookie recipes are particularly important examples. I chose them because everyone understands what the names mean. The two examples are merely stand-ins for any software component. I can't use realistic ones since they are either obscure and too hard to explain, or nonexistent because of the absence of revenue collection solutions like the one herein addressed. If the apparent triviality of the cookie recipe troubles you, substitute whatever software component interests your company right now. Perhaps a platform-independent Window class, a calendar management function for insurance companies, a module that computes loan payments for banks, a marketable piece of clip art, a newsletter, or what have you. The cookie recipe is only an instance of a class that could be very broad indeed once the underlying economic problem has been addressed.

ShrinkWrapIt is simply an electronic analog of the paper and cellophane packaging of the traditional software store. It is not the only such product on the market, but competes with many others on the basis of function, quality, and price. Such products provide an external surface, a "free part," that customers can decorate with product advertising and the mandatory explanation of prices, terms, and conditions. They also provide an internal place, the "value bearing part," where subject-matter experts like Aunt Nellie can construct their wares. They could also provide a menu of pretested, ready-to-use terms and conditions algorithms (TCAs) that Aunt Nellie could understand and use easily, and a slot for the crucial product ID number that will be issued during product registration. Such "product wrapper" products could even drive the product registration process electronically and hide the complexity of the product development and registration process to the point that ordinary folk like Aunt Nellie could play.

The complexity is, of course, higher for those who build products from first principles. In addition to the difficulties familiar to all programmers, we can no longer hide behind disclaimers of warranty for customers harmed by an errant activity of our code. The truly revolutionary part of this approach is that it enables customers in a way they never have been before, so let's look at how this works in more detail.

In choosing between competing shrinkwrap products for her recipe object, Aunt Nellie compared functionality, quality, prices, terms, and conditions. She did this just as she might comparison shop for tangible goods, by browsing free parts of the competing vendor's products, just as she might browse competing flour packages in the grocery store. Just as she might buy a small sample to bake a trial batch if the advertisement captures her interest, she might try out an object, thus incurring invocation fees, to see how well their value-bearing parts meet her needs.

Such shrinkwrap products contain two pairs of query/commit instructions. The outermost pair charges the customer in favor of Aunt Nellie's account. The second pair is enclosed in the previous pair and charges Aunt Nellie in favor of the shrinkwrap vendor.

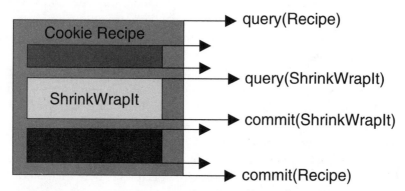

Figure 7.5 It may seem backwards that the shrinkwrap in this figure is *inside* the product, not outside. Containment here refers not to the physical relationship between wrapper and contents, but to their relationship in the structure of a production tree. Aunt Nellie is the owner of the recipe object at the top of the hierarchy. The owner of ShrinkWrapIt is a supplier one level down. If ShrinkWrapIt is itself composed of smaller objects (as it well might be), additional query/commit pairs would be involved in exactly the relationship shown above.

7.12 TERMS AND CONDITIONS ALGORITHMS

A key step in the product registration process involves providing a computer program, called the *Terms and Conditions Algorithm,* or TCA for short. The financial institution will use this algorithm in its billing cycle to convert the invocation counts generated by the query/commit instructions into monetary amounts.

Each month the financial institution receives an encrypted record of the query/commit calls made by its users. Each query/commit record contains the pid (product ID) of the product they represent. The pid identifies the TCA to be used to compute a monetary amount to be credited to the owner. The nesting of the query/commit pairs reflects the nesting of multigranular products, so the immediately enclosing record determines whose accounts will be debited.

Since TCAs are algorithms, the relationship betweeen invocation records and monetary amounts can be as simple or as complex as a marketeer's ingenuity might devise, so long as it complies to the rules of fair trade discussed earlier. One simple algorithm is the pay-to-own terms and conditions that dominate the market today. The TCA simply computes a fixed amount the first time a product is used by a customer and nothing forever thereafter. Incremental charging, where a much smaller fixed amount is charged for each query/commit pair is even simpler. Any point on the pay now versus pay later continuum could be implemented in this manner. For example, pay-if-you-like it involves delaying the fixed pay-to-own fee to allow a free evaluation period.

7.13 FINANCIAL INSTITUTIONS

No single institution could muster enough computers and communication capacity to keep up with (potentially) millions of users. Each invocation record involves very small overheads to the end-user's computer, since the record involves no more than appending a few arguments to a cache. The intrinsic costs of accomplishing this, particularly with a tamper-resistant cache based on silicon that is tightly coupled with the CPU (which is desirable for tamper resistance reasons anyway), are comparable to the overhead of a subroutine call which does essentially the same thing.

The records generated by each call are uploaded to a computer at a financial institution where each record must be processed by a terms and conditions algorithm whose overhead will in general be substantially larger. How would a financial institution keep up with calls each month from millions of advanced personal computers, each comparable to the fastest computers the financial institution might buy?

Put this way, the question answers itself. In principle, there is no way to keep up since records are generated faster than they can be processed. However, the world doesn't operate in principle. Those computers are being used to do work, not to generate invocation records. Nor do people work all the time. Although it is technically possible for pranksters to overload the processing, their contractual agreement with the financial institution would provide language to inhibit such behavior. Computationally intensive terms and conditions algorithms could either be outlawed outright, or inhibited by financial charges.

Furthermore, it is neither desirable nor necessary that all of the work be done centrally. Visa and MasterCard don't process credit card transactions centrally. They franchise thousands of member banks to process most of the volume locally, using telecommunications technology to coordinate their activities.

Finally, "financial institution" is not a synonym for "bank." They are data processing businesses, consolidating thousands of micro transactions each day into larger monthly amounts that traditional banks can handle efficiently. That is, most of the transaction volume involves credits and debits between accounts of the data processing institution. Traditional banks need only be involved when an account owner asks to withdraw or deposit cash.

Another implication is that the financial institutions need not be large. They could be small businesses operated by ordinary citizens, perhaps even from home. The technical requirements are modest: a network interface (which could be a telephone and modem), a computer, and a franchising relationship with an umbrella organization comparable to that between Visa and MasterCard and its member banks. The franchiser would provide the training, support, software, and so forth. It would enforce rules against behaviors that might detract from its good name, similar to the rules that fast food franchises impose on their members.

A final implication is that the financial institution need not be external. For example, major corporations are very unlikely to agree to any system that publishes detailed records of what objects their employees have been using off premises, since this information can be exceedingly revealing of what the corporation is planning. However, they might be perfectly willing to operate their own financial services server locally. This server would process invocation records locally, transforming the sensitive invocation-level information into aggregated financial amounts. These should raise no more concern than check or credit card information which they're familiar with, and receptive to, already.

7.14 SUMMARY

The design presented in this chapter involves a strange combination of good news and bad news. The design is based on a remarkably simple observation, that charging for software based on invocation is more feasible than basing revenue on acquisition of goods that can be transported at the speed of light.

The good news is that the techno-centric issues are themselves exceedingly simple. The entire infrastructure is based on nothing more complicated than pair query/commit instructions that just write their arguments into a cache. Although this solution is sufficient for *objects* (data encapsulated in software), a deployed system (such as that of Wave) would also provide encryption capabilities for protecting the unencapsulated data that so much of the industry is concerned with today.

The bad news is that techno-centric issues are such a tiny part of what's involved here. Even when there are clear incentives for doing things differently, people and institutions resist change, particularly when the old ways seem to be working. Convincing silicon manufacturers to provide the infrastructure in silicon could easily take years. Persuading successful software suppliers to build software that can be distributed freely over networks could easily be as difficult, especially since existing distribution infrastructures (software stores) encourage today's focus on large unigranular software. And even though the most likely outcome of full deployment of such a system is to put far more power into software consumers' hands than we've ever experienced before, the most likely short-term reaction is to resist it as a departure from "personal computing" and an invasion of end-user's privacy.

I am clearly an optimist that this approach will be deployed as we make a transition from the industrial age to the information age. However, the historical examples provided in this book suggest that pessimism should prevail both in the short term and with respect to the software institutions that are successful today. Successful agrarian-age enterprises, such as cottage industry gunsmiths, didn't adjust to industrial-age practices. They fought them with total devotion, and they did so successfully, until the cottage industry was itself replaced by new enterprises that were never successful under the old approach.

Chapter 8

Conclusions

This concluding chapter is a personal retrospective on the path that led me to the controversial and decidedly nonmainstream approaches that I've advocated in this book.

8.1 THE ELECTRONIC FRONTIER

This book takes Kobayashi's first two waves for granted and focuses on his third, the integration of computers and communications into the everyday affairs of mankind. In this worldview, "electronic frontier" is not just a flowery metaphor, no more than "The Wild West" once was. It is the name of a human-centric process that people are engaging in all around us today. Calling it a metaphor denies the reality of this new world to the electronic authors and programmers, those who want to earn a living by building electronic goods for sale. For us, the electronic frontier is no metaphor but the world we're beginning to call home.

However, this frontier is still in formation. Since World War II, computing was almost exclusively a techno-centric affair. Power resided almost exclusively in the hands of the technical sophisticates able to meet computers on their own terms. But with the introduction of personal computers, accelerating radically with the explosive growth of the worldwide web, a shift to a human-centric worldview is occurring.

The technical minutia that once made computers and communications the domain of a technical elite have been increasingly eliminated. Instead of being objects of enthusiasm, envy, frustration, and awe, computers are vanishing, disappearing into the woodwork to become just part of the plumbing. Having the latest and greatest computer no longer compels the interest that it once did. Computers are becoming smaller, more powerful, and increasingly easy to use, disappearing from ordinary people's sight as well-behaved plumbing should. They are becoming transparent *windows* through which **ordinary people** can communicate, coordinate, cooperate, and compete as members of an advanced socio-economic community.

This opening of the frontier to more than a technical elite spells the end of this elite's long dominance. It means the slow emergence of an entirely new and utterly different world, a process that I've called the taming of the electronic frontier.

8.2 A PERSONAL RETROSPECTIVE

This book has described a controversial vision of how the electronic frontier will develop over time. Other prominent visionaries, people as capable and insightful as Esther Dyson, Mitch Kapor, and John Perry Barlow of the Electronic Frontier Foundation and many others, have entirely different ideas.

I will not elaborate these visions here. Rather I'll use this closing chapter to describe the underlying assumptions of these competing, and for the most part dominant, visions. To distinguish this dominant model from the one discussed in this book, I've called these competing visions "the established paradigm."

The Electronic Frontier, as with the other frontiers of history, has its own indigenous culture. As in all other frontiers, these indigenous tribes are being encroached upon by foreign invaders. I hope that neither of these communities will take affront if I use here the names that the two sides use for each other. I will call the indigenous tribe the Nerds and the foreign invaders, the Newbies.

I am a long-standing member of the Nerd tribe. I speak its language fluently and fully share its view of the world. To me, boot is a verb as often as a noun. Terms like operating system, unix, and even IP/TCP are so much a part of my vocabulary, and my worldview, that I barely notice the confusion such terms cause to outsiders. Even though I try hard to avoid using them, the students in my classes complain that I regularly confound them with terms that are utterly incomprehensible to them. I am regularly brought up short by the realization that most students enter these classes with no more comprehension of my language than the Boers understood Bantu or the Spanish understood Cherokee.

Of course, the breakdowns run both ways. I am just as unable to understand the language Newbies use to describe their computer problems to me. They lack not only the vocabulary, but the very worldview, to communicate computer problems to me. "I typed in what you told me and it didn't work. What did I do wrong?" Or every Unix expert's all-time favorite, "I couldn't find the ANY key," in response to the ubiquitous prompt, "Hit any key to continue."

I am as immersed in the worldview of this freedom-loving indigenous culture as any cypherpunk.[1] My tribe's deepest core values are embodied in the term 'privacy'. As I've described here, my intellect believes that we'd be better off by reengaging privacy as an electronic property issue and not as a privacy issue.

At a deeper emotional level I immediately respond to this term, privacy, as strongly as anyone. I am referring, of course, to the FBI's wiretap bill that was passed into law in 1995, the Clipper Chip of the Clinton administration proposal that thankfully has not been passed, and the possibility that strong public key cryptography may soon be outlawed or that anti-export restriction on cryptography be maintained. I even maintain a web page for privacy-related issues, *Big Brother is Watching* web at http://www.virtualschool. edu/mon/BigBrother.html.

My tribe is now at the pinnacle of its power and influence. I am told that Bill Gates, the founder of Microsoft, is the wealthiest man in the U.S. right now. Newt Gingrich, another member of my tribe, is the most powerful individual in Congress. Bill Clinton and Al Gore's advocacy of the National Information Infrastructure qualifies them, at least in outsiders' eyes, as members of the same tribe. Of course, their enthusiasm for violently interferring with other people's interests in the War on Drugs, the War on Tobacco, the Waco and Weaver massacres, and the Clipper Chip proposal, shows that they also hold very different views from other parts of this tribe.

Regardless of the obvious differences between tribal factions in the preceding examples, they all share a common understanding of what network infrastructures are for. In this established paradigm, these are *information* infrastructures, not *commerce* infrastructures. As I developed this distinction in the preface, networks are information infrastructures like the pony express and very unlike commerce infrastructures like the railroads. Underlying this distinction is the tacit conviction that the problematics of owning goods-made-of-bits are so deep-seated that robust commerce in electronic objects is forever beyond our reach.

Furthermore, there are powerful examples that seem to show that the electronic frontier will continue as it has in the past, organized primarily by the communitarian short-range binding forces of indigenous tribes and not by the commercial, longer-range but far weaker, forces of advanced industrial cultures. The internet, the personal computer revolution, unix, and so forth provide powerful exemplars of what noncommercial social orders can accomplish. These examples seem to argue that the electronic frontier will

---

[1] Cypherpunks is a netnews group that emphasizes radical privacy-enhancing techniques such as anonymous remailer technology, cryptography, and direct political action. Cypherpunks can be thought of as the Nerd tribe's warrior caste.

trigger a resurgence of the communitarian ideal that might someday connect warring factions around the entire globe into a single "global village." In this new world, commerce would be relegated firmly to the margins, a refuge for the minority who "don't quite get it."

In this view, light-speed communication infrastructures extend the range of the short-range forces that were previously confined to face-to-face ranges, restricting their influence to communities, villages, companies, and tribes.

8.3 WILL THINGS REALLY DEVELOP THIS WAY?

My crystal ball is far too cloudy to say that things won't develop precisely along these lines. As the other paradigm shifts I've presented show, paradigm shifts are not easy to predict and even harder to influence.

The path that led me to this unconventional view is that my career in software engineering has given me an ample opportunity to see that the established paradigm holds no solutions to the software crisis. Of course, this is hardly compelling to those who have never had an opportunity experience the software crisis first hand.

Sometimes I wonder what today's world might be like if the retrograde motion of Mars hadn't been obvious to every untrained peasant's naked eye. If this breakdown in the established astronomical model had been obvious only through the Ptolemaic elite's telescopes, might not they have overlooked it forever, exactly as we are overlooking the Software ICs crisis today?

My experiences with Objective-C and Software-IC showed that object technology had made it technically quite feasible to build software components that other people could and would reuse. Chapter 5 also showed that it was technically feasible to specify, document, and test components with advanced specification and testing tools comparable to the inspection gauges of the industrial revolution. However, this also showed that while all this is technically feasible already, it is not economically feasible, and economic feasibility is crucial to moving any such initiative ahead.

If one horse can generate one horsepower, how much can two horses generate? The answer ranges widely, from less than zero to more than two, depending on factors external to the horses. If they're harnessed to pull in opposite directions, the answer can be zero. If they're left unharnessed to do what comes natural to stallions and mares, the outcome may even be more than the sum of the parts. The outcome depends on factors external to the horses, on what I've referred to as binding forces in this book. Binding forces coordinate individuals into a social entity. The power of this larger entity can be greater than, or less than, the sum of its parts.

Other examples are all around us. In addition to syphilis and the power of the invaders' guns, a major reason for the displacement of the Indians was their inability to build lasting alliances between tribes. This made it possible for the Spaniards, and later the French and English, to form alliances with one tribe that bore some grudge against another. This strategy was carried out to great effect by Cortes in his conquest of the Aztecs and in the French and Indian wars. The Indians' effective power to ward off conquest was far less than the sum of the parts because invaders could harness them to work against one another.

We have all witnessed this phenomenon in firms, those communities that gather around conference tables on the job. When there's no longer-range force to coordinate the participants' efforts, the output of a committee is often a powerful, stress-inducing logjams whose effective horsepower is zero. Although I can point to equally disappointing outcomes in industry, academic faculty meetings are great places to watch stress-producing stagnation in progress, with little discernible result.

One further example from this world should suffice to demonstrate how external binding forces can radically influence the power of the whole. When I was a child, I had a part-time job as a bag-boy for the Pigley Wigley grocery store in Lake City, South Carolina. Although I've no reason to believe that Grover Iric, the store manager, had formal management training, he wrote a memo that I've used ever since as a masterpiece of organizational thinking. Contrast Grover Iric's memo with the parking policy at my own university, which greets newly arriving taxpayers, visitors, students, and staff with "Faculty and Staff Only" signs on the most desirable slots, just at the point when industry has been absorbing and acting on powerful lessons about the importance of empowering their customers:

Dear Employees:

1. The parking places next to the door are reserved for Pigley Wigley customers.
2. Employees should park toward the other end of the parking lot.
3. The slots at the very end of the lot are reserved for Pigley Wigley management.

Thank you; Pigley Wigley Management

Clearly, this little gem of organizational insight originated from the commercial forces that so many academics disapprove of. However, the absence of these same forces perpetrates unconscionable attitudes toward students and taxpayers that are likely to cripple universities forever.

My reading of history suggests that societies organized around longer-range commercial forces tend to fare better than those organized around the shorter-range face-to-face forces of the communitarian ideal. However, history also shows the overwhelming diversity in how people have organized their affairs. This makes the question of whether things will actually go as I have outlined, and more importantly, when, something that my crystal ball is utterly unable to discern.

The ancient Aztec and Mayan cultures revolved around a form of massive blood sacrifice on the steps of their pyramids that would have apalled even the devoted Nazi. Yet this culture prospered for ages until it was brought down by the Spaniards. Every human culture I can think of right now, including the Egyptians, Greeks, Romans, Nazis, Indians, and my own Confederate States, at one time or another used forced slavery to harness other people into what was then regarded as a cohesive whole.

Tribal communities still exist today, in North and South America, Africa, Australia, and Asia, that have existed almost unchanged since antiquity. Long-range commerce and its correlary, individualism, is entirely subsidiary in these cultures to the tribal communitarian ideal. Even Christianity, arguably the largest and most successful social innovation ever, amounts to a communitarian binding force in which commerce plays no overtly explicit role. Although the collapse of Russia sent the socialist central planning fashion into something of a decline, emergent nations still turn to monarchs, dictators, and central planning bureaus to coordinate their internal affairs.

The point is that, no matter how implausible or repugnant, many different binding forces have been tried and have proved sufficiently successful to persevere for very long times. This is only one of the reasons why it is far from certain that the commercial forces that developed from the industrial revolution will soon encroach onto the electronic frontier. A second, and much smaller, reason is that this encroachment isn't even possible in the absence of a technical commerce infrastructure like the one discussed in Chapter 7.

My crystal ball is much better at looking back than looking forward. When I turn it this way, it shows an endless succession of communitarian cultures that are soon forced to retreat as soon as they're confronted by cultures in which commercial forces are strong. I have tried to show in this book that this same confrontation may be heading our way, and have shown one way in which the invading Newbies might surmount the natural barrier created by goods made of bits.

Do I expect the Nerd community to change its ways and adapt this new approach voluntarily? No, I most certainly do not. The history of the American Indians, the Williamsburg Gunsmith, and Thomas Kuhn's discussion of paradigm shifts in science, show that paradigms are very deeply held indeed. I anticipate, and fear, that the electronic frontier will be tamed as the Wild West was tamed, by the displacement of an establishment by some outside group that could never have been successful under the old paradigm.

Index

A

Abstract data type, 113, 114, 131
Account management, 170
Ackoff, Russell, 63
Aquinas, 48
Ada, 77
Adams, John, 118
Agassiz, Jean Louis, 63
Alchemists, 48
Alchemy, 49, 66
America Online, 168
American Precision Machinery Museum, 115
American Society of Composers, Authors and Publishers, 149
Analysis, 60
    Analysis/design methodology, 53, 60
ANSI, 58
Apple Computer, 6
Appliance-level object, 104
Architectural diversity, 94
Architecture, compositional, 79
Architecture, software, 74
Argument list, 175
Aristarchus, 48
Aristotelian cosmology, 48
    Aristotle, 48, 55
Armament industry, 114
Armory practice, 115
ARPAnet, 12
ArrayBasedCollection, 137
Artistic social order, 22
ASCAP, 149
Assembly, 53
    Assembly industry, 66
Assembly intensive, 127
    Assembly technology, 93
assert.h, 132
    Assertion checking, 132
Astronomy, 48
Astronomy crisis, 49
Austrian economics, 26
Aztecs, 195

B

Bantu, 192
Barker, Joel, 60

Barlow, John Perry, 11, 192
Being Digital, 9
Big Brother, 192
Binding force, 29
Binding technology, 78, 81
Bionomics, 33
Bits and atoms, 8
Black box testing, 134
Blanc, Honore, 118
Blanchard Pattern Lathe, 120, 121, 130, 138
Blanchard, Thomas, 120, 124
Blank canvas, 13
Blanket license, 149
Block-level object, 96, 106
BMI, 149
Bolt and die, 59
Bomford, 126
Booch, Grady, 51
Breadth vs. depth, 8, 10
Breech Loading Rifle of 1819, 120
Broadcast Musicians Incorporated, 149
Broadcast technology, 149
Brooks, Frederick, 45, 48, 65, 111
Browser, 70
Build to order, 127
Bureau of ordnance, 117
Burke, James, 48
Burns, Robert, 111
Buy, sell, and own, 12

C

C, 106
C++, 40, 58, 59, 74, 108, 131
Cache, 174, 175
Calculation debate, 28
Calhoun, John C., 121, 126
Capitalistic ideal, 17
Card-level object, 96
CASE (Computer Aided Software Engineering), 52, 60, 127
Cash transactions, privacy of, 170
Central planning, 34
    Centrally planned bureaucracy, 12
Cerf, Vint, 11
Change, 2
Changeability, 68
Cheops, 82
Cherokee, 192
Chicken vs. egg, 68, 91
    paradox, 63
CHILL, 83
Chip-level object, 96, 105
Choctaw, 17
CID, 170

Clarke, Arthur C., 49
Class, 138
Clinton Administration, 6, 192, 193
Clip art, 15, 185
Clipper Chip, 14, 192
Closed system, 105
Cobol, 59
Coding, 60
Coercion, 27
Collection, 137
Colonial Williamsburg Museum, 115
Colt, Samuel, 124
Commerce, 35, 42
Commerce infrastructure, 168
Commercial exchange transaction, 30, 44,
   47
Commercial ideal, 196
Communication infrastructure, 168
Communication with financial institution,
   175
Communitarian, 29, 193
   ideal, 17, 196
Compile-time type checking, 131
Complexity, 22, 48, 68
   software, 51
Compositional architecture, 79, 81, 95
Compuserve, 79, 168
Computer industry, 40
   science, 54, 77
   software, 37
Computers as communication devices, 6
   as places, not things, 7
   as windows, 7, 191
Computers, Communication and Mankind
   (CC&M), 2, 157, 191
Concrete data type, 113, 114, 131, 132
Concurrent computer architecture, 70
Configuration control system, 53
Conformity, 68
Conservation of mass, 14, 30, 31, 44, 150,
   155, 170
Constitution, U.S., 149
Contractual law, 170
Conway's Law, 81
Copernican Revolution, 8, 48
   Copernicus 49, 55, 78
Copy protection, 152, 157
CopyFree Software, 155
Copyright, 149, 150
Coroutine, 69, 104
Cortes, 195
Cosmology, 54
Cosmology, heliocentric, 49
   Heliocentric system, 48
Cottage Industry Craftsmanship, 56

Cotton gin, 118
Cox's Corollary, 81
Credit card application, 170
Credit cards, privacy of, 170
Crystalline spheres, 48
Cultural belief, 93
Customer ID, 170
Cyberspace, 14
Cypherpunk, 192

D

Dark Ages, 66
DARPA, 51
Data encryption standard, 180
Denver International Airport, 21, 32, 37
Department of Defense, 11, 51
DES, 180
Descartes, 49
Design, 60
Digitalk, 98
Dijkstra, Edsgar, 54
Dillon, Marshall, 14
Distributed computing, 70
Dominant paradigm, 81
Drexler, Alan, 162
Dynabook, 12
Dyson, Esther, 192

E

Ease of duplication, 68
Economic agent, 27
Economic feasibility, 194
Economics, 72, 143
Eiffel programming language: 59, 131
Electronic frontier, 10, 15, 191
Electronic Frontier Foundation: 14, 192
Electronic goods, 12, 22, 43, 63, 72, 111, 150,
   170
Electronic objects, 193
Electronic pencil, 23, 39
Electronic property, 12, 14, 15
Elite priesthood, 55
Ely Whitney Museum, 115
Encapsulation, 85, 92
   Encapsulation mechanism, 81
Encryption, 152, 178
Englebart, Douglas, 92
Epicycle, 48, 49
Equilibrium economics, 27
Escobar, Pablo, 12
Essence and accident, 45
   Essence vs. accidents of software
   engineering, 65
Established incumbent, 49

Established paradigm, 53, 54, 81, 111, 192
European culture, 11
Evolution, 22
Exception handling, 70, 133
Exchange transaction, 144
Exemplar, 61
Experimental observation, 55

F

Fabrication, 53
Fabrication and assembly, 84
  Fabrication vs. assembly, 41, 60, 77
Fabrication technology, 93
Fairness to consumer, 183
Fairness to producer, 183
FBI Wiretap Bill, 192
Fetzer, H., 55
Fiedler, Edgar R., 45
Figure-ground distinction, 91
Financial institution, 170, 187
Finite message machine, 82
Formal method, 53
Formal proof of correctness, 130
Fraud, 183, 184
Free part, 181
French and Indian Wars, 195

G

Galileo, 49, 78
Gate-level object, 96, 106
Gates, Bill, 92, 193
Gauge, inspection, 113
GEnie, 79
Geographic proximity, 31
George Mason University, 9, 28
Gibbs, W. Wayt, 21
Gingrich, Newt, 193
Goldberg, Adele, 78
Gonzales, Barbara, 36
Gonzales, Derek (Po-Ye-Mu), 37
Gonzales, Maria, 37
Goods made of atoms, 14, 37
Goods made of bits: 15, 17
Goods, electronic, 72
Gore, Al, 193
Granularity, levels of, 79
Graphical user interface, 70
Gunsmoke (TV Series), 14

H

Hall, John, 115, 119, 124, 126
Hardware engineering, 66
Hardware store, 112

Harpers Ferry, 120, 121, 126
  Harpers Ferry Armory, 115
Hassle factor, 149, 150
Hayek, Freidrich, 26
Heavyweight process, 104
Heterogeneity, 77
High, Jack, 28
Hoare, C. A. R., 55
Home page, 10
Huberman, B. A., 155
Human-centric, 2, 8, 22, 183, 191
Humphrey, Watts, 53
Hydraulic system, 27
Hypermedia, 4
Hypertext, 16

I

IBM 360, 3
IBM 7090, 3
IBM 7094, 3
ICPak, 42
Implementation
  component, 140
  hierarchy, 137
  technology, 130
  tool, 113
Indian, 17, 196
  culture, 11
  pottery, 35
  Way, 35, 37
Indigeneous
  culture, 192
  encounters on the electronic frontier, 17
Industrial Age, 1
  goods, 13
  jobs, 16
  metaphor, 27, 33
Industrial Revolution, 21, 53, 63, 66, 111,
  114, 188
  heroes of, 126
Industrial Social Order, 22
Information Age, 1
  goods, 13, 143
Information economy, 13
Information pump, 162
Information Revolution, 8
Inheritance, 138
Inheritance hierarchy, 137
Inline procedure, 106
Inspection gauge, 123, 126, 128, 132, 138,
  194
Inspection team, 53
Intangibility imperative, 54, 58
Intangible abstraction, 55, 75

Intangible goods, 2, 63
Intel, 41, 43
Interchangeable component, 53
    part, 115, 116
    software component, 88
Interface Convention, 81
Internet, 8, 35, 168, 193
Invisibility, 68
Invocation metering: 174, 183
Invocation-based revenue collection, 168
Iric, Grover, 196

J

Japan Electronics Industry Development
    Association, 155
Jefferson, Thomas, 118, 124
JEIDA, 155, 157
Job security, 56
Jobs, Steve, 92
Joubert, Joseph, 46

K

Kapor, Mitch, 14, 92, 192
Kawahara, Masaji, 155
Kay, Alan, 12, 74, 92
Kepler, 49, 78
Kobayashi, Koji, 2, 6, 16, 149, 157, 160,
    191
Kuhn, Thomas, 61, 196

L

Lake City, South Carolina, 196
Language, 59
Large granularity object, 174
Large granularity solution, 152
Lavoie, Don, 28
Law enforcement, 183
Laws of man, 150, 170
LectroNubbin, 174
Lee, Roswell, 118, 119, 120, 123, 124
License sever, 180
Lightweight multitasking, 69
Lightweight process, 104
ListBasedCollection, 137
Livesay, Harold, 118
Logic, 60
    and reason, 48
Lone Ranger, 46
Long-range force, 17, 27, 31, 193
Lotus, 92
Lovato, Charles, 35
Lumberjack, 27
Luther, Martin, 49

M

Macintosh, 79, 93
MacTCP, 5
Management information system, 13
Maniac II, 3
Market mechanism, 29
Market process, 30, 54, 162
Mass spectrogram, 79
MasterCard, 187
Meaning-making, 2
Meyer, Bertrand, 131
Micrometer, pocket, 124
Microsoft, 92, 193
Microsoft Word, 40
Miller, Mark, 43, 155, 162
Mises, Ludwig, 28
Molecular orbital calculation, 3
Mopping-up operation, 62
Mori, Ryoichi, 155, 168
Multigranular goods, 167, 177, 180
Multimedia, 70
Music industry, 149
Musket, 115
Mythical man-month, 45

N

Napoleon Bonaparte, 118
NASDAC, 178
National Semiconductor Corporation, 178
NATO Software Engineering Conference,
    45
Natural Law, 170
Negroponte, Nicholas, 9
Nelson, Ted, 92, 155
Nerd, 192
    tribe, 17
Nested record, 175
Newbie, 17, 192
Newton, Isaac, 29, 79
Nintendo, 160
Nippon Electric Corporation, 2
Nisus, 40
Nobel Prize, 28
Normal science, 61
North, Simeon, 124
Novobilski, Andrew, 106
NSA, 170

O

Object, 137
    Objects all the way down, 78
Object technology, 16, 64, 74, 143, 164, 194
Object-oriented design, 51

Object-oriented programming language, 78, 92, 138
Object-oriented technology, 52, 95
Objective-C, 41, 59, 74, 106, 127, 131, 194
Obstacle, 67
Open system, 105
Operating System, 93
OrderedCollection, 137
Ordnance department, 123
Organon, 48
Out of the crisis, 144
Oxymoron, 77

P

Packet switching, 12
Paradigm, 62, 196
Paradigm shift, 7, 49, 53, 60, 126, 161, 194
Paradigm, established, 47, 51
Paradigm, patching, 48
ParcPlace Systems, 98
Pay per copy, 73, 149, 160
Pay per use, 149
Pay to acquire, 167
PDP 11/70, 98
PDP 8/I, 98
Pencil tree, 23, 30, 33, 144
Personal computer revolution, 193
Personal retrospective, 192
Philosopher's Stone, 48, 66
Physical conservation law, 14
Physical universe, 68
PID, 175, 183
Pigley Wigley Grocery, 196
Piracy, 15
Planck, Max: 2, 62
Plato, 48, 55
Platonic ideal, 66
Plumbing, 6, 16, 52, 53, 73
Pony Express, 168
Postcondition, 133
Potlatches, 11
Pottery, 35
Power, 57
Preconditon, 133
Prefabricated software component, 53, 54, 60, 77
Price, 177, 181
Price signaling, 27, 30
Pricing, software, 152
Privacy, 15, 170, 192
Problematics of owning electronic property, 193
Procedural programming language, 56
Process industry, 66
Process Maturity Certification, 52

Process vs. product, 89, 95
Process-centric, 58, 116
Process-level object, 96
Prodigy, 168
Producer hierarchy, 170
Product ID, 175, 183
Product registration, 170, 185
Product-centric, 58, 60, 116
  world, 72
Programming language, 93, 127
  and methodologies, 143
Proximity, Law of, 92
Ptolemaic cosmology, 48, 66
  elite, 194
  ideal, 54
Ptolemy, Claudius, 48
Public key cryptography, 192
Pueblo Indian, 35
Puzzle solving, 61

Q

Query and commit instruction, 175, 186
Query/Commit record, 181
Queue, 137
Quicken, 170

R

Rape seed farmer, 27
  oil, 31
Rapid prototyping, 60
Rasp, 125
Read, Leonard E., 26
Reengineering the organization, 34
Registration process, 183
Retrograde motion of Mars, 194
Reusable software component, 15, 32, 43, 60, 167, 194
Revenue, 71, 168
Revenue collection infrastructure, 155
Revenue infrastructure, 170
Revolutions, time course, 124
Robust commerce, 193
Rothschild, Michael, 34
Royalties, 152
Rules of Fair Trade, 183
Rutkowski, Tony, 8

S

San Ildefonoso Pueblo, 36, 144
Santa Fe Institute, 35
Scarcity as basis for commerce, 30
Scarcity pulse, 28
Schopenhauer, Arthur, 63

Scientific American, 21
Scientific revolution, 48, 49, 54, 60, 61, 66, 111
  structure of, 61
Scully, John, 6
Sega, 160
SEI (Software Engineering Institute), 51
Semaphore, 137
Semiconductor industry, 66
Shareware, 152
Shaw, George Bernard, 67
Short-range forces, 17, 193
ShrinkWrapIt, 185
Silicon chip, 85, 96
Silicon foundry, 77
Silicon Valley, 40
Silver Bullet, 45, 48, 49, 65, 66, 75, 111
  Law Enforcement Products, 46
Single threadedness, 68
Smalltalk, 59, 74, 93, 98, 131
Smith, Merrit Roe, 115, 120, 126
Smithsonian American History Museum, 115, 118, 121, 123
Social binding force, 29
Social Order, 21, 27, 38, 144
Software, architecture, 74, 94
Software, best selling, 79
Software complexity, 50
Software components, 113
  marketplace, 69
Software crisis, 21, 32, 45, 144
Software development tool, 143
Software engineering, 16, 23, 27, 32, 45, 54, 77, 111, 127
  computer aided, 52
Software industrial revolution, 53, 63, 88, 117
Software lifecycle and lifespiral, 86
Software reliability, 16
Software reuse, 60
Software werewolf, 32
Software-IC, 32, 41, 69, 98, 168
Software-ICs, 194
Soviet Union, collapse of, 26, 27
Specialization of labor, 53, 69, 73, 85, 88
Specialized labor hierarchy, 54
Specification
  component, 140
  hierarchy, 138
  technology, 130
  tool, 55
  in mature engineering domain, 140
  testing and language, 112
  testing technology, 71, 111, 132
  testing tools, 113, 128, 194

SPECmark, 5
Spirituality, 35
Spokeshave, 125
Sprague, Peter, 178
Springfield Armory, 115, 118
Standard process, 114
Standard product: 114
Standardized parts, 53
Standish Group International, 167
Stepstone, 41, 42, 127, 152, 168
Sterling, Bruce, 14
Stiles, Chris, 17
Stroustrup, Bjarne, 106
Structure of production, 21, 27, 44, 144
Stubblefield, James, 119
Sumser, J. R., 39
Superdistribution, 155, 168
Superdistribution and object technology, 164
Syllogism, 48
System 12, 82

T

Taming the electronic frontier, 9, 13
Taming the Wild West, 11
Tamper resistance, 157, 178, 183
Tamper resistant repository, 175
Tangibility imperative, 56, 58
Tangible goods, 2, 22, 41, 43, 63, 150
Task-level object, 96, 104
TCA, 181, 185, 186
Techno-centric, 2, 8, 16, 17, 183, 188
Technosocial law, 170
Telecomputing industry, 12
  technology, 16
Telephone switching system, 82
Terms and conditions, 175, 177, 181
  algorithm, 181, 185, 186
Thread of control, 68, 104
Time constant, 62
Timesharing, 69
Tolerance, 54
Tools of certainty, 123, 130
Tools of risk, 123, 130
Treating the symptoms, not the disease, 143

U

Unigranular goods, 167
University of Chicago, 3
Unix, 92
Unscrupulous behavior, 184
Use of knowledge in society, 26
Usenet, 12, 79

User registration, 170
Useright, 149, 150

V

Value bearing part, 181
VDM (Vienna Definition Method), 56
Virtual reality, 4
Visa, 187
Viscosity, 1
VLSI, 92
   design, 82

W

Waco Massacre, 193
War on Drugs, 193
Waterfall model, 86
Wave systems, 178
Weaver, 193
Web, 15
Wells Fargo, 14
Werewolf, 75
   software, 111
White box testing, 133

White collar productivity, 13
Whitney, Eli, 118, 124
Williamsburg gunsmith, 196
Winchester Silvertip Hollow Point, 46
Winer, Dave, 17
Winsock, 5
Word processor, 23, 39
World of the mind, 55
World Wide Web, 10, 16
Wozniak, Steve, 92
WYSIWYG, 4

X

X3J16, 58
Xanadu, 155
Xerox Parc, 12, 98

Y

Yahoo, 46

Z

Z, 56